Intelligent Mars II:

Code of the Craters

Arthur Raymond Beaubien

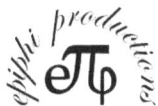

epiphi productions

Intelligent Mars II: Code of the Craters

ISBN 9780994032119

epiphi productions
Ottawa, Ontario
Canada

www.epiphiproductions.ca

Library and Archives Canada Cataloguing in Publication

Beaubien, Arthur Raymond, 1942-, author
    Intelligent Mars II : code of the craters / Arthur Raymond Beaubien.

ISBN 978-0-9940321-1-9 (paperback)

    1. Martian craters. 2. Mars (Planet)--Surface. I. Title.

QB641.B433 2016   559.9'23   C2016-906614-2

Cover design © by Arthur R. Beaubien. Mars map on cover courtesy USGS Astrogeology.

Dedication

To those who dare to be themselves

Contents

v   Preface

1   Chapter 1. Ancient Prime Meridians, Prime Latitudes &
                Degree Systems
27 Chapter 2. Square Craters
55 Chapter 3. Pentagon Craters
72 Chapter 4. Hexagon Craters
88 Chapter 5. Octagon Craters
95 Chapter 6. Martian Craters Come in Standard Sizes
104 Chapter 7. The Denning and Savich Craters
114 Chapter 8. Janssen's Wheel
131 Chapter 9. The Elorza Alignments
140 Chapter 10. The 6E3S Crater
154 Chapter 11. The Eye of Sharonov
167 Chapter 12. Issedon Tholus and the Ayacucho Crater
187 Chapter 13. Craters on the Mountains
206 Chapter 14. Craters as Teachers

Note: The term *sacred distance formula* used in this book refers to sacred geometry formulae of the form $(i/j)mnr$, $(i/j)(1/m)(1/n)r$ or $(i/j)(m/n)r$ where i and j are integers, r is the equatorial (R) or northern polar (R') radius of Mars, and m and n have the value of 1, φ, π, e, $\sqrt{2}$, $\sqrt{3}$, or $\sqrt{5}$. The simplest formulae are either R or R' in which all the other variables are set equal to 1. Distances are in terms of km. When used to refer to a coordinate value, the term *sacred formula* refers to the same formulae forms as used for distances but without the 'r' variable. Coordinates are in terms of degrees or radians.

Preface

The *Intelligent Mars* series is devoted to uncovering and analyzing the covert artificiality of the Martian landscape. *Intelligent Mars I* * was focused mainly on demonstrating how the mountains are arranged according to sacred geometry. *Intelligent Mars II* is a detailed study of the craters on Mars. It also presents the discovery of a multiplicity of ancient Martian coordinate systems.

There are thousands of craters on Mars, so many in fact, that we have been led to believe that this was always a dead planet exposed to billions of years of bombardment by asteroids and comets leaving huge holes in the landscape. The first indication that some of the craters might actually be artificial in nature rather than naturally formed was the discovery of survey craters discussed in *Intelligent Mars I*. I was content for a long period of time to simply assume that these were the only craters that were artificial, created for the special purpose of marking out the location of the centres of important mountains. Eventually, however, I realized that a number of craters were not simply circular holes, but contained straight line segments, notches, steps and other deviations in their perimeters and interiors. Such aberrations pointed to the existence of a code embedded in the various structural features of these craters which could only have come into existence by intelligent engineering.

At first I reasoned that crater locations and crater abnormalities might mark out some sort of grid which could be used by overhead spacecraft to determine their positional coordinates but I could find no evidence to support this. I also experimented with the possibility that several craters might lie on long straight lines which pointed to important sites. This also proved to be a dead end. Then I noticed that many craters had perimeters which could be fit to certain geometric shapes such as squares and other polygons. An amazing finding was that polygon-shaped craters are sized according to music intervals found in the chromatic scale used by our own musicians on planet earth. The use of polygon shapes and the sizing to music intervals suggested that several craters might actually be encoding a form of sacred geometry. But the biggest key to breaking the code of the craters came in the remarkable discovery of multiple prime meridians, prime latitudes and degree systems used by the ancient Martian civilization. These were employed by the architects to determine where to position craters and their anomalies so that crater features and fitted geometric shapes would have coordinates which reflected sacred geometry primarily in the form of the irrational constants of φ, π, e, √2, √3 and √5, and in the integers representing the angle sizes found in the pentagram, equilateral triangle and square.

So on one level, the code of the craters embeds the message of sacred geometry. But the Martian use of sacred geometry must have had an even deeper message of its own. This would appear to be the proclamation of a spirituality that values harmony with a Divine Creator who created the universe and themselves out of sacred numbers and geometric shapes. The Martian civilization wanted their home planet to resonate with the act of creation and hence with the Being that created it. Their architectural creations were to assist them with union with the Divine much like Yoga techniques and meditation are intended to do for its practitioners. I believe that this is the chief message that we can learn and should take from our study of ancient Mars.

Arthur R. Beaubien

* *Intelligent Mars I: Sacred Geometry of the Mountains. Did Da Vinci Know? Arthur Raymond Beaubien. Epiphi Productions, Ottawa, Ontario, Canada. 2015.*

Acknowledgements

The author wishes to express his gratitude to Gayle Peterson and his daughter Kery for giving an abundance of meaning to his life. He also wishes to thank George A. Neville for his invaluable enduring friendship, suggestions and support, and to Liana Manole, Gabrielle Markvorsen, and Keith Bailey for their generous gift of time and many helpful comments in their review of this manuscript.

Cover: *The Sharonov Crater has the shape of an eye which "looks over" the mountain architecture thousands of kilometers away. The map segment of Mars is courtesy of USGS Astrogeology.*

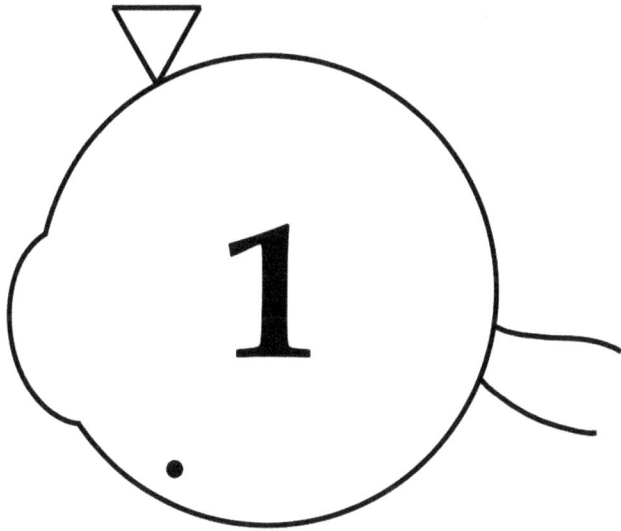

Ancient Prime Meridians, Prime Latitudes & Degree Systems

I t would be extremely impractical to start the discussion on the craters of Mars without first providing the reader with an understanding of the suprasystem of organization which I have discovered to exist on Mars. I have found that almost every site that has been artificially constructed is referenced in some way to this system which integrates these structures in such a way that they create a vast global monument composed of mountains, craters and other landforms. In fact, the numbers generated with reference to this system reveal the artificiality of sites so clearly that it becomes impossible to rationally deny that an advanced intelligence has been responsible for their construction.

The suprasystem that I am referring to is the collection of ancient prime meridians, prime latitudes and degree systems which I have uncovered after many years of studying the planet's topography. My knowledge of this suprasystem has come about by a combination of pure luck, careful examination of the Martian topography and countless measurements to test various hypotheses. In the chapter entitled "Lessons in Longitudes and Latitudes" from *Intelligent Mars I*, I suggested that the longitude for the survey centre of Elysium Mons was probably the best candidate for the original prime meridian (PM) on Mars. I also concluded that the latitude of Arsia Mons provided a second prime latitude (PL) in addition to the equator and that a 720 degree coordinate system may have been used by the Martian civilization. Now, after studying many more craters and artifacts on Mars, I have come up with a series of startling discoveries which suggest that while my original choice for the prime meridian still holds true, the Martians have used several other prime meridians as well. Based on the history of our own planet, we would be tempted to think that the explanation for this is that the prime meridian may have been changed several times over the very long history of Martian occupation. However, a much more likely explanation is that several prime meridians were used simultaneously prior to site construction to create important numbers in the coordinates of the sites or of their components. The same can be said for the several Martian degree systems which I am about to reveal.

A Serendipitous Discovery

My story starts with a chance discovery while exploring the significance of the Sharonov Crater (301.5358° E 26.9969° N). If you look closely at the crater (Fig. 1.1), you will notice that there is a light-coloured narrow linear structure lying just inside the western side of the crater. In the middle of this structure there is a small extension on its eastern side. The other

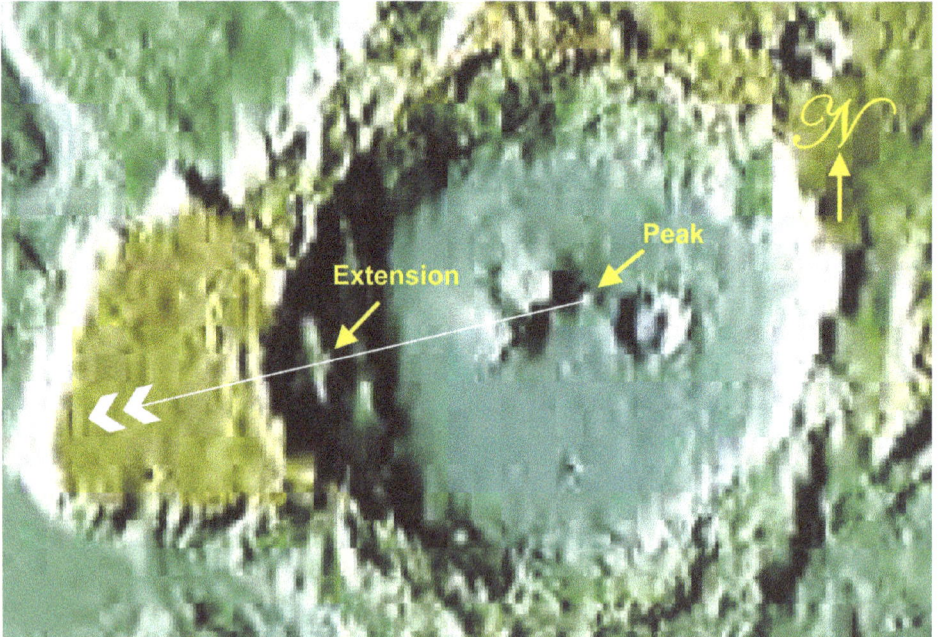

Fig. 1.1: *Line drawn from a "Peak" in the Sharonov Crater so that it passes through the middle of an eastern extension of a narrow light-coloured structure inside the western portion of the crater. The line has a counterclockwise bearing angle very close to 103° and eventually passes through the Fesenkov Crater (see Fig. 1.2). USGS Astrogeology.*

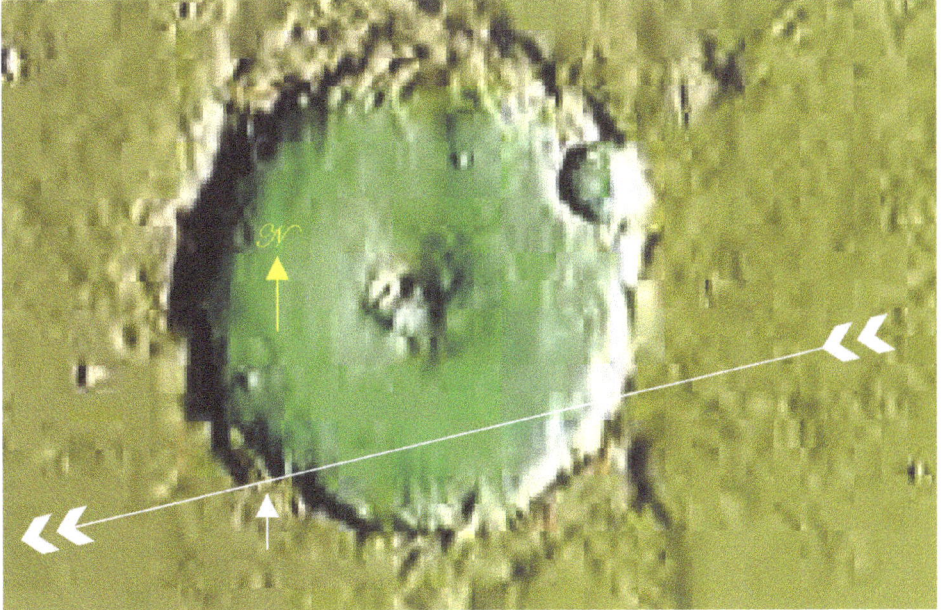

Fig. 1.2: *Line continued from the Sharonov Crater shown in Fig. 1.1. This line passes through the Fesenkov Crater coinciding with the southern edge of the southwestern notch in the crater's perimeter (see white arrow at lower left). USGS Astrogeology.*

feature of the crater that I wish to draw your attention to is a tiny white structure (labeled in Fig. 1.1 as "Peak") located in the central area of the crater. During my initial attempts to figure out how the Sharonov Crater related to the other sites on Mars, I noticed that if I passed a straight line from the Sharonov Peak through the middle of the eastern extension of the linear structure and extended the line more than 1500 km westwards, the line eventually aligned to the edge of an outward notch in the southwest perimeter of the Fesenkov Crater (273.4705° E 21.6321° N, see Fig. 1.2). Not being content to stop here, I was curious to see if the line would cross any other important sites. It eventually led me to an area on the other side of the planet which lay well south of Elysium Mons and below the equator. There I encountered a rather dark area with bright parts scattered here and there. What was unusual about the area was that it did not seem to fit into the surrounding terrain, and had a completely different texture to it.

"Is this a dagger which I see before me....?"

I studied the area for some time, zooming in and out to try to get some sense of what I was seeing. Then suddenly a pattern emerged, a pattern which was the last thing that I would have expected to find on Mars. A sword....? That's what it looked like at first.... some sort of sword which pointed southwards with a gentle curvature towards the east. But what looked like a handle at the northern end seemed too big in relation to the blade of the instrument for it to be a sword. No, it was much more likely to be a dagger than a sword (See Figs. 1.3, 1.4). Now what would the image of a dagger be doing on the surface of Mars? I thought for a moment and wondered if a sword or dagger image had ever been used to represent the Earth's prime meridian on a map. An Internet search quickly turned up what I was looking for, only it was not a representation of the current prime meridian passing through Greenwich, England. It was a Merovingian styled sword overlaid on a map of France with the midline of the sword blade marking the prime meridian passing through Paris[1]. This drawing appeared on the cover of a novel called the *Circuit* deposited in the Bibliothèque de Versailles in 1971. It was written by Philippe de Chérisey, one of the principle persons behind the claim of the existence of a secret society called the Priory of Sion which many people today believe was a hoax. So incredibly, a Martian civilization existing perhaps more than 3 billion years ago was probably doing exactly the same thing - using a sword or dagger to mark out a prime meridian for the planet. Was this an original idea reborn here on Earth? Or has the tradition been past on in some way from the planet Mars?

The line from the Sharonov Crater and the notch on the Fesenkov Crater passes through the hilt of the dagger. The longitude of the midline of the northern part of the hilt between its east and west sides turns out to be only 9 seconds of a degree more than 1° E of the longitude of the survey centre for Elysium Mons. This suggests that the midline plays a prime meridian role. This is somewhat analogous to the Paris Prime Meridian (Paris PM) being 2° 20' 14.025" east of the Greenwich Prime Meridian. Which came first on Mars is anybody's guess at this point. One possibility is that the Dagger Midline Prime Meridian preceded the construction of both Elysium Mons and the great mountains on the Tharsis Rise, and was used as a reference to base the locations of these huge structures. Another good possibility is that the Dagger Midline and Elysium Mons prime meridian candidates were created simultaneously to serve different purposes. Since numbers appear to be extremely important to the Martian architects, a displacement of 1 degree would create different numbers for the longitudinal values of any particular site.

I also measured the bearing angles (angles from due north) of linear sections of the dagger hilt outline. There were 5 sections for which there was adequate information on which to base the fit of a straight line. The bearing angles (negative numbers are clockwise angles) of all these sections are evenly divisible by 3, and four of them are evenly divisible by 6. The number 12 used for the northern sides of the handle is a sacred

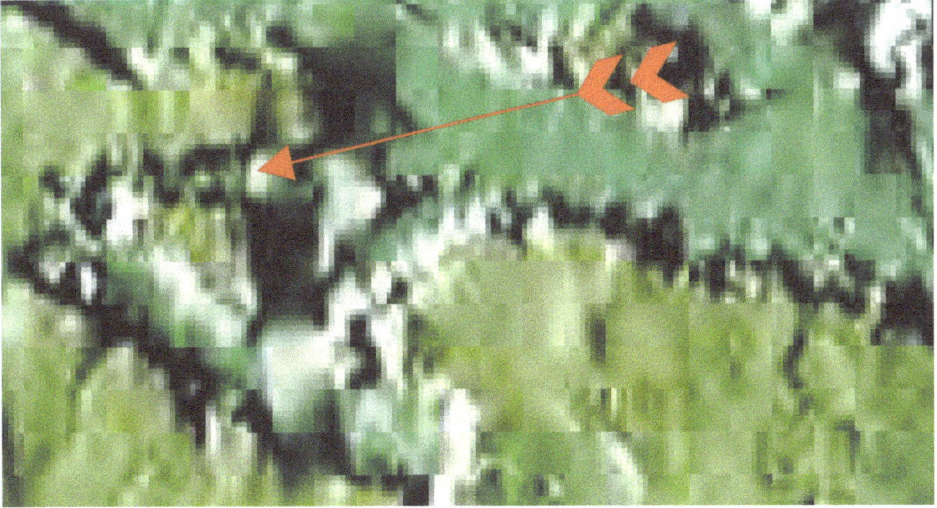

Fig. 1.3: *The red arrow is the line from the Sharonov Crater extended far beyond the Fesenkov Crater to a region below the equator not far from the longitude of Elysium Mons. Note the region of light and dark areas centred on the red arrowhead. This region is outlined in Fig. 1.4 below. USGS Astrogeology.*

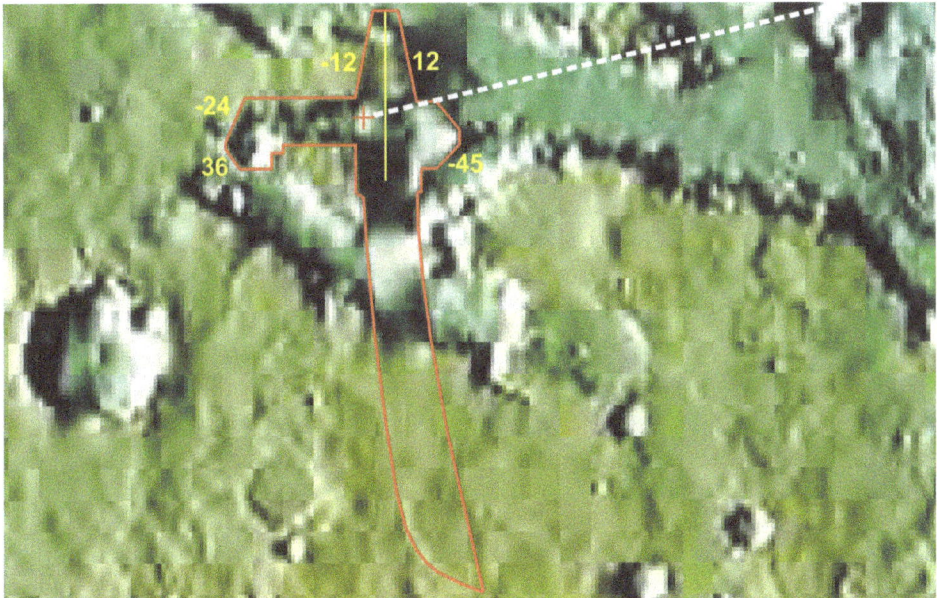

Fig. 1.4: *Approximate outline of a dagger found on the Martian surface. The dashed white line is the line extended from the Sharonov Crater. The yellow numbers on the outside of the dagger outline are the bearing angles of various straight line sections. The vertical yellow line is positioned on the midline of the northern part of the dagger hilt and marks the Dagger Midline PM which is 1° E of the Elysium Mons PM. The red cross placed on the white area in the middle region of the dagger hilt marks the site of the Dagger Peak PM (see discussion later on in the chapter). USGS Astrogeology.*

number as is discussed in Chapter 12 of *Intelligent Mars I*. It divides evenly into the 24 and 36 values. The number 36 is the size in degrees of the angle of a star point of a pentagram. An angle of 45° likely refers to a square or rectangle since it is one-half the value of their 90° angles. The use of these numbers for bearing angle values suggests that the dagger was an integral part of the sacred geometry of the major architectural sites on Mars and may therefore have been created at about the same time as the other sites.

The Sharonov Tower Prime Meridian

Being led to the dagger from the Sharonov Crater half a planet away was a stroke of pure luck. Because the midline of the dagger handle lay almost exactly 1° E of the Elysium Mons Prime Meridian, I had every reason to believe that this site also marked a prime meridian. But my luck did not stop there. I discovered that there are very deep connections between the dagger site and the Sharonov Crater which led me there, as well as with the Elysium Mons site. When I returned to examine the Sharonov Crater in more detail, I decided to measure the coordinates of a tiny white structure in the interior of the crater near its southern outer perimeter (Fig 1.5). This structure is similar in dimension to the Sharonov Peak in Fig. 1.1 and seemed to represent some sort of a peak or tower. I refer to it as the Sharonov Tower to distinguish it from the Sharonov Peak. Incredibly, its brightest pixel centred at a longitude of 301.1743° E was found to be only about 15 seconds of a degree further than the integer of 154 degrees from the Elysium Mons Prime Meridian. Hence, it would be almost exactly 153 degrees from the suggested prime meridian marked by the midline of the dagger hilt. Since these numbers are both very close to integer values, the purpose of the Sharonov Tower was most likely to act as an alternate prime meridian in integer sync with both the Elysium Mons PM and the Dagger Midline PM to create different numbers than would occur with either of the other 2 prime meridians. These numbers would carry their own significance. A good example of this is the longitude of the Pavonis Mons Caldera. With Elysium Mons as prime meridian, the longitude of this site has a value of 100° E. With the Sharonov Tower as Prime Meridian, the longitude would be 54° W. The number 54 is one-half the value of the interior angles of a pentagon or of the angles between 2 adjacent star points in a pentagram. Hence, the number 54 would be a reference to all the symbolism associated with these geometric figures that would not be achieved with the number 100. The difference of 153° between the Sharonov Tower and the Dagger Midline PM seems unremarkable at first until you realize that 153° is the supplementary angle of 27° (i.e., 180° - 153° = 27°). The angle of 27° is 1/4

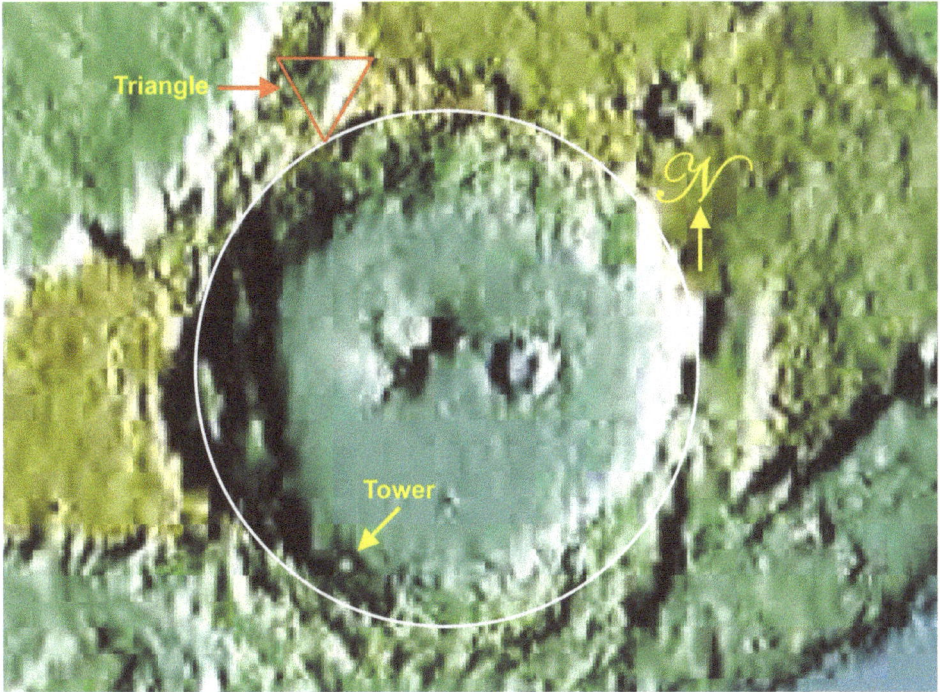

Fig. 1.5: *Picture of the Sharonov Crater showing the locations of the Sharonov Tower and the Sharonov Triangle prime meridians. The large white circle is fitted to the outer perimeter of the crater. USGS Astrogeology.*

the size of the interior angles of a pentagon or of the angles between 2 adjacent star points in a pentagram. It is also 3^3 which emphasizes the number 3. But we are not yet done with the Sharonov Crater.

The Sharonov Triangle Prime Meridian

An even more remarkable finding concerning the Sharonov Crater was the discovery of an equilateral triangle in the landscape just outside the northwest part of the crater (Fig. 1.5). The southwestern side of the triangle is well defined by a trench which is marked by 3 red arrows in Fig. 1.6. Most of the southeastern side is also easily visible as a trench and is marked by 2 red arrows in Fig. 1.6 (bottom right). The northern side is defined by colour transitions along part of its course which denote changes in elevation. These are marked by a pair of red arrows in the upper part of Fig. 1.6. It is possible that the northern side is actually connected to an arrow shaft which would make the triangle an arrowhead instead of simply a triangle, but this not well enough defined to ascertain with certainty, so I have elected to simply call this anomaly a triangle.

My fit of this landform to a triangle is shown in Fig. 1.5. Since the

Fig. 1.6: *Triangular marking just outside the northwest perimeter of the Sharonov Crater. The southwest side (3 red arrows bottom left) and southeast side (2 arrows bottom right) of the triangle appear to be linear trenches while the northern side (pair of arrows at top) is indicated by transitions in colour. The triangle is equilateral having 3 equal sides and 3 internal angles of 60°. One of its vertices points due south. USGS Astrogeology.*

southern vertex of the triangle touches the perimeter of the crater, I decided to call this triangle the Sharonov Triangle. The point of contact of the triangle with the perimeter of the crater has a longitude of 301.1052° E which is not an integer distance from either the longitude of the survey centre of Elysium Mons (147.1701° E) or that of the Dagger Midline PM (148.1727° E). The deviations from integer values by 0.0649 and 0.0675 degrees, which are close to 4 minutes of a degree, are too large to be accounted for by inaccuracies in my measuring techniques. Yet the triangle has the unmistakable appearance of a prime meridian marker with its close association with the perimeter of the Sharonov Crater. The mystery appears to be solved when you look at the longitudes of some sites other than those which are synchronized by integer or integer and 1/2 displacements to the survey longitude of Elysium Mons. If you recall from Chapter 10 in *Intelligent Mars I* there was a group of 9 sites (Group 1) whose longitudes were in integer or integer plus 1/2 sync with one another. However, no site could be identified as a prime meridian for this group. Let's now look at the coordinate displacements of these same 9 sites from the longitude of the tip of the Sharonov Triangle (Table 1.1). In these and subsequent measurements in this chapter, I allow a tolerance of ±0.0333 degrees = ±2 minutes of a degree for the designation of "integer" or "integer and 1/2". In Table 1.1, all of these sites meet this definition in their longitude displacement from the Sharonov Triangle.

The most interesting longitude displacement in Table 1.1 is that of the Pavonis Mons survey centre since, like the Pavonis Mons Caldera, it is 54° W of a good prime meridian candidate. The value of 54 is 1/2 the size of the angle between the star points of a pentagram. Because the pentagram encodes the golden ratio $\varphi = 1.6180$ in many ways, the presence of a longitude of 54° W for both the caldera and survey centre of Pavonis Mons suggests that this mountain may have been strongly associated with the

Table 1.1: *Site longitudes from the Sharonov Triangle PM.*

Site	Longitude (°W Sharonov Triangle PM)	Longitude (Rounded to 1 Decimal Place)
Apollinaris Mons	126.9759	127.0
Ascraeus Mons Caldera	45.4766	45.5
AscSC2 Crater	51.9866	52.0
Biblis Tholus	64.9922	65.0
Hecates Tholus Caldera	150.9973	151.0
Pavonis Mons	54.0010	54.0
Pentagram Pyramid	69.4745	69.5
Ulysses Tholus Caldera	62.4995	62.5
Ulysses Tholus N Crater	62.5000	62.5

sacred geometry of φ. Further evidence of this is the discovery reported in *Intelligent Mars I* that the latitude of the survey centre of Pavonis Mons is $\varphi°$ N in planetographic coordinates, and that the latitude of the caldera of Pavonis Mons is $(\varphi/\pi)°$ N in planetocentric coordinates. It was also reported in *Intelligent Mars I* that there is a rectangular area coming off the northeast corner of the caldera of Pavonis Mons which has a bearing angle of approximately 9 degrees in the clockwise direction (See Fig. 1.11 below). The similarity of the astrological symbol for Mars to the configuration of the caldera joined to a rectangle led me to postulate that this configuration may have been used as the original symbol for Mars, and that the tip of the arrow was later added to the end of the rectangle. Now, considering all the associations of Pavonis Mons with the pentagram and φ, the intriguing possibility arises that the same symbol may also have been used for φ itself. The rectangle may have eventually been rotated to line up vertically at the bottom of the circle to represent the golden ratio (Fig. 1.7). Note that one of the variants of the symbol for the golden ratio is very similar to the depiction of a planet with its axis of rotation passing through it. Also noteworthy is that the bearing angle of the rectangle is about 9° clockwise. This ties it to the pentagram and therefore, the golden mean, since 9° is 1/4 of the angle of 36° at each pentagram star point.

Thus it seems extremely likely that the triangle touching the Sharonov Crater was used as a prime meridian in its own right, and it likely coexisted with the Sharonov Tower and several other prime meridian markers simultaneously. This is not as unusual as it might seem, since the recognition of more than one prime meridian at the same time has occurred many times in the past on Earth. For example, although the

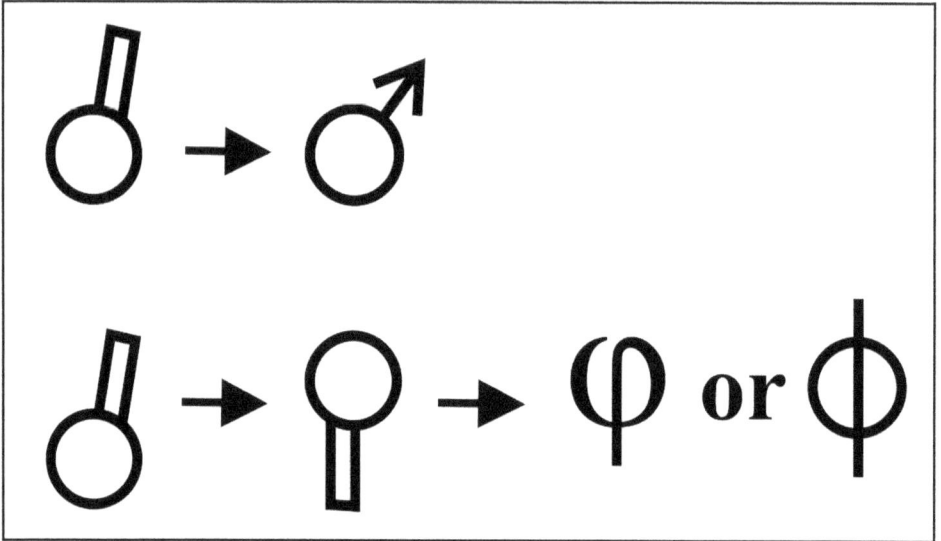

Fig. 1.7: *Pavonis Mons "symbol" composed of its caldera and a rectangular area adjacent to the caldera. With the passage of time, this "symbol" may have been converted into the astrological symbol for Mars (upper row) and the symbol for the golden ratio (lower row).*

Greenwich Prime Meridian was accepted internationally in 1884, the French carried on with the Paris Prime Meridian for several decades afterwards. Sea charts often used 2 or more prime meridians such as those on Danish ships which measured degrees from Greenwich, Copenhagen and Paris. However, the purpose on Mars appears to be more connected with creating sacred geometry numbers rather than for navigation.

The Big Degree

It would appear that with the discovery of 4 good candidates for prime meridians we should now be able to explain everything about the coordinate systems used by the Martian architects. But the story is far from over yet. I stumbled onto the next piece of the puzzle at the time I was compiling data on hexagonal-shaped craters (see Chapter 4) in the Northern Hemisphere. When I came across the Bamburg Crater (356.9528° E, 39.6497° N), I noticed that it's width seemed close to 1 degree of longitude. I decided to measure the crater's width a little more carefully. When I took a measurement, I found it to be about 1.16 longitude degrees, which is definitely larger than 1 degree. I was about to give up on this line of thinking when a crazy idea went through my mind. What if the crater actually did represent 1 degree of longitude, but the degrees were of a different size than what you would get in a 360 degree system? To test my hypothesis, I divided 360 degrees by 1.16 and got 310.3 degrees as an answer. Of course nobody would set up a coordinate system based on 310.3 degrees. So maybe there was some error in my measurement, or else, my hypothesis was indeed crazy. So what is close to 310.3 that would allow for a practical coordinate system? The number 300 would not work very well since the distance from the equator to a north or south pole would be measured as 75 degrees which has impractical subdivisions. The same goes for 310 degrees. How about 320 degrees? That would make each degree in the new system equal to 1 1/8 or 1.125 ordinary degrees. Not too crazy, especially when you realize that each quarter of the circle could then be a sum of 10 equal divisions instead of 9 in the 360 degree system. Let me explain that a bit. In the 360 degree system there are 90 degrees from the equator to one of the poles. Hence, the 90 degrees can be divided into 9 equal divisions of 10 degrees each. It could also be divided into 10 equal divisions of 9 degrees each, but the 9 degree units would be very clumsy to further subdivide so it is never done. In the 320 degree system there would be 80 degrees between the equator and one of the poles. This would allow for 8 units of 10 degrees each, or for 10 units of 8 degrees each. The latter way of grouping the degrees offers an easy way of further subdividing the units since you are now into a binary system, making each further subdivision a simple division by 2. This is even more practical than 10 degree units since the factor 5 is not evenly divisible by the factor 2. Thus a grouping of degrees in units of 8 is likely to be the most plausible choice for a 320 degree system. I decided to call a degree in this tentative theoretical system a *big degree* in order to distinguish it from the degree size we normally use.

The Dagger Peak Prime Meridian

My first test of the novel degree system came when I attempted to calculate the number of big degrees between the Sharonov Tower and both Elysium Mons and the midline of the dagger handle. As you may recall, assuming integer displacements, the distance in ordinary degrees was 153 degrees to the dagger handle, and 154 degrees to Elysium Mons. Dividing each of these displacements by 1.125 yields 136.8889 big degrees for Elysium Mons and... 136.0000 big degrees for the middle of the dagger handle! This was the first indication that I might just be on to something here. It not only pointed out the possibility that big degrees were used on Mars, but also offered further support to the use of the midline of the dagger handle as a prime meridian. But what about the Sharonov Triangle? Might there be another location on the dagger that would do for the Sharonov Triangle what the vertical centre of the handle does for the Sharonov Tower location? Careful examination of the dagger site indeed revealed a great candidate for such a point. If you go back to Fig. 1.4 you will see a red cross placed on a white-coloured area just west of the vertical centre of the dagger handle at the level of the crossbar of the dagger hilt. The centre of the 4 brightest pixels in this area has a longitude of 148.1052°E which turns out to be 153.0000 degrees west of the Triangle of Sharonov and translates nicely into 136.0000 big degrees! With this result, I now had a 5th good candidate for a prime meridian since this bright area was associated with the dagger, a construct which was likely to have been used to mark a prime meridian. I called this newest addition to the prime meridian family the Dagger Peak PM. You will also notice in Fig. 1.4 that I have drawn the line from the Sharonov Crater so that it terminates exactly at the location of the red cross. This is theoretical rather than exact since the resolution of the map did not permit me to determine the course of the line to such accuracy. I can only say that it passes somewhere through the dagger hilt.

Now I was faced with the problem of discovering which of my 5 prime meridian candidates were actually used for the coordinates of sites on Mars, and whether big degrees were really put to use or... were my findings so far just wild speculation? But first, as a practical note, I needed some sort of system to easily distinguish between the 2 types of degrees. It needed to be simple, concise and easy to recognize to reduce confusion. To this end I devised the method of using a double degree symbol to denote big degrees. Thus, for example, 36°° would mean 36 of the big degrees in the 320 degree system and 36° would mean 36 conventional degrees in the 360 degree system. So be on the lookout for the double degree symbol. It will not be a typographical error (I hope).

Latitude degrees

As a final comment on the big degree, I later found out that my inspiration based on the longitude width of the Bamburg Crater was pure blind luck. Although, as you shall see, there is very strong evidence for the existence of the big degree, my route to its discovery was erroneous. As discussed in Chapter 4, hexagon-shaped craters were sized by the Martian architects in terms of latitude degrees, not longitude degrees. The distance covered by a degree of longitude varies according to the latitude since meridian lines converge from the equator to the poles. However, a degree of latitude in planetocentric coordinates is constant in length (59.2747 km) regardless of where it is measured on the planet so it makes a convenient standard of measurement. It is equivalent to the length of a degree of longitude measured at the equator. It appears that the Martian architects used this unit of measurement for sizing polygon craters. Thus the width of the

Bamburg Crater should be considered to be about 0.9 latitude degrees rather than 1.16 longitude degrees. So once again science moves erratically forward by a lucky error rather than leaping ahead by a brilliant theory.

The Pentagon Pyramid

The existence of a 320 degree system remained somewhat dubious for me until I made another extraordinary discovery. I unexpectedly found the Rosetta Stone of coordinate system confirmation for both the 360 and the 320 degree systems. It also confirmed the use of the Dagger Peak as a prime meridian. This happened while I was searching the territory southeast of the Elysium group of mountains. I suddenly noticed an object which appeared to be a perfect regular pentagon (Fig. 1.8). As it seemed to be an aboveground structure and definitely not a crater, I named the object the Pentagon Pyramid. It is not to be confused with the Pentagram Pyramid located some 4000 kilometres away to the east. Like the Pentagram Pyramid, the Pentagon Pyramid was no trivial object, having a side length of about 25.16 kilometres which makes it approximately 139 times the 180.7 meter side length of the US Pentagon in Arlington, Virginia. A substantial portion of the eastern half of the pyramid appears to have suffered some damage since its outer edge is not well defined, especially on the southeast side. This is reminiscent of

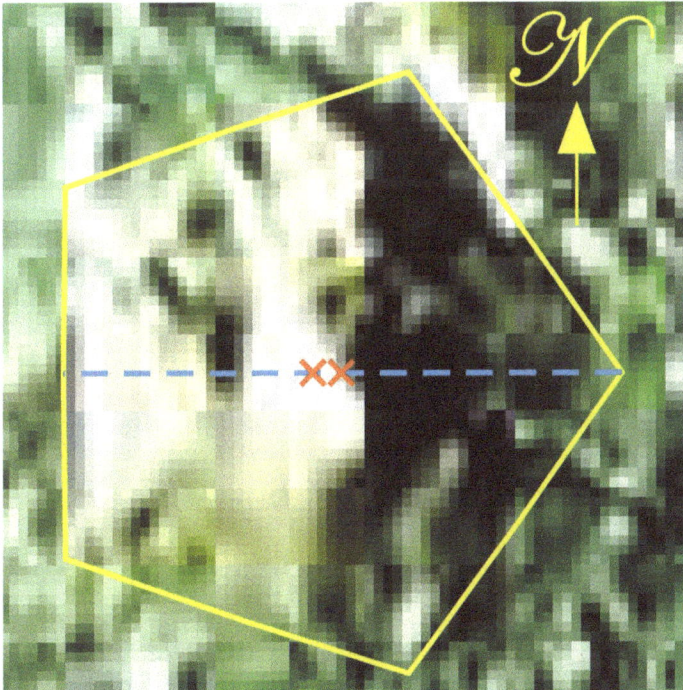

Fig. 1.8: *Pentagon Pyramid located southeast of Elysium Mons. Each side is about 25 km long. The left red "x" marks the pentagon centre while the "x" on the right marks the centre between the west side of the pyramid and its eastern vertex. The latter centre has the matching coordinates 12°° E (Dagger Peak PM) and 12° N. USGS Astrogeology.*

the Cydonia face and the D&M Pyramid on the other side of the planet where both structures appear damaged on their eastern halves. The distance from the centre to any vertex is about 21.4 km or very close to one-thousandth the planetary equatorial circumference of 21,339 km! The diameter of a circle drawn around its perimeter is therefore equal to about 42.8 km or very close to 0.72 latitude degrees. Since the Pentagram Pyramid was estimated to have a diameter of 0.5 latitude degrees, the diameter of the Pentagon Pyramid is about 1.44 times the size of the Pentagram Pyramid, making it a truly colossal structure.

The centre of this particular object lies at coordinates 161.5738° E 12.0000° N. When I used the various candidates for the prime meridian, the longitudes which were calculated turned out to be rather uninteresting numbers. The only number that came close to being an integer or an integer and one-half was the value obtained with the Dagger Peak PM, namely a value of 13.4686° E. However, if I used the coordinates of the midpoint of the distance between the west side of the pyramid and the eastern vertex, I discovered that with the Dagger Peak PM, its longitude was 13.5031° E. Now that's much more interesting. The latitude was still 12.0000° N since this midpoint has the same latitude as the centre of the pentagon due to the fact that the eastern vertex points due east and the western side of the pyramid lies in an exact north-south direction. When I converted the longitude of this midpoint to the big degrees of the 320 degree system, I got 12.0028°° E. I just about fell out of my chair. Here was an obviously very important site whose east-west midpoint had matching coordinates of 12° N and 12°° E! All you had to do was use the regular degrees for latitude and the big degrees for longitude.

This elegant location was hugely important since it substantiated (a) the Dagger Peak as a PM, (b) the existence of a 320 degree system and (c) that both the 360 and 320 degree systems were used simultaneously. This did not necessarily exclude the use of the other candidates for the prime meridian, but it certainly solidified the existence of a prime meridian at the longitude of the Dagger Peak. The emphasis on the number 12 is also very interesting since it shows that this number was very important to the Martians. It was found as the bearing angles of the 2 sides of the northern portion of the dagger hilt (Fig. 1.4). It was also found as a power of 2 for the determination of the Martian kilometer and the Martian meter from the equatorial radius of Mars (see Chapter 12 of *Intelligent Mars I*).

The Sacred Degree

Just when I thought that I had finished uncovering all the major coordinate systems used on Mars, with 5 possible prime meridians and 2 systems for dividing the planet into degrees (actually 4 systems if you consider that 720 and 640 degree systems might have also been used, converting half degrees into integers), I found that the width of another hexagonal-shaped crater was close to 1.25 regular degrees of longitude (crater E596 in my numbering system at 30.8066° E 2.1811° N, Table 4.1, Chapter 4). Note that because this crater is close to the equator, the difference in distance covered by longitude and latitude degrees is slight. This stimulated my thinking once again. Could there be yet another system to divide the planet into degrees? I played around with a few numbers and decided that 1.25 degrees did indeed give a credible system. Dividing 360 degrees by 1.25 gives 288 degrees. At first this seemed like a rather clumsy number, but when I started to divide it by a sequence of 2's I realized that this would give a series of numbers which have been considered very sacred by ancient cultures. Thus you get 288, 144, 72, 36, 18 and 9. Sound familiar? The values of 36 and 72 are the sizes in regular degrees of the angles in the star point triangles of the pentagram. In this system, the northern and southern hemispheres would each have 72 degrees of latitude which could be divided into 12 divisions of 6 degrees, or 6 divisions of 12 degrees. The number 12 is another important sacred number as discussed above. Alternatively, 72 degrees could be divided into 9 divisions of 8 degrees each. The 8 degree subunits contain a binary number of degrees which repeatedly divide by 2. Besides being practical, the presence of so many numbers regarded as sacred by multiple past civilizations may have been enough to motivate a 288 degree system. Even the number 9 can be further subdivided into 3 groups of 3 which would certainly fit with the emphasis on the number 3 that is found with all the triangular arrangements of the mountains of Mars. However, this would probably be found to be less than practical. Because of all the sacred numbers that crop up when 288 degrees are divided into various subunits, I decided to call the degrees of this new system *sacred degrees.*

To test out the sacred 288 degree system, I started measuring the longitudes and latitudes of important sites with reference to the 5 prime meridians in terms of the 3 different degree sizes. The first sites I tested were the sites that provide the greatest substantiation for the Elysium Mons Prime Meridian. To start with, I examined the centre of the Pavonis Mons Caldera which is 100.00° E of Elysium Mons. It is also 99.00° E and 88.00°° E of the Dagger Midline Prime Meridian, and 54.00° W and 48.00°° W of the Sharonov Tower PM. When I measured the longitude in terms of sacred degrees, I found that this site to be 80.00°°° E of the Elysium Mons PM. Notice that I use 2 consecutive degree symbols to represent the 1.125 degree size (big degrees), and 3 consecutive degree symbols to represent the 1.25 degree size (sacred degrees). Ascraeus Mons is 11.28° N, 10.03°° N and 9.02°°° N. It is also 95.49°° E of the Dagger Midline PM, 40.51°° W of the Sharonov Triangle PM and 36.52°°° W of the Sharonov Tower PM. The southern peak of Tharsis Tholus is 122.52° E and 98.02°°° E (Elysium Mons PM), 121.52° E and 108.02°° E (Dagger Midline PM), and 31.48° W and 27.98°° W (Sharonov Tower PM). The most interesting coordinate of these is the 108°° E, even though it is in terms of big degrees, because 108 is the number of regular degrees between the star points of a pentagram. This is a good instance of how the use of alternate prime meridians and degree systems can be used to generate the numbers which are considered very sacred or symbolic. Biblis Tholus is 88.01° E of the Dagger Peak PM, and 64.99° W and 51.99°°° W of the Sharonov Triangle PM. The crater on the northern side of Ulysses Tholus is 50.00°°° W of the Sharonov Triangle PM. It is also 3.14°° N (i.e., π°° N). One of the craters (see Fig.13.13) on Hecates Tholus is a remarkable e°°° E of the Dagger Midline PM [e = 2.7183 is the base of the natural log].

Mountain calderas seem to be especially in sync with the sacred degree system. Thus the northeastern caldera of Olympus Mons is 80.03° E and 64.02°°° E of the Elysium Mons PM. It is also 15.00°°° N of the equator. The central caldera of Olympus Mons is 59.50°°° W of the Sharonov Triangle PM. The central caldera of Ascraeus Mons is 86.02°°° E of the Dagger Peak PM. The caldera on the top of Uranius Tholus is 91.51°°° E of the Dagger Midline PM, whereas the circular flat top itself of Uranius Tholus is 91.50°°° E of the Dagger Peak PM. The old caldera of Apollinaris Mons, as reconstructed from its perimeter to the southeast, is centred at 101.51°°° W of the Sharonov Tower PM. The Biblis Tholus Caldera has 2 centres. The southern centre is 70.48°°° E of the Dagger Peak PM and 51.98°°° W of the Sharonov Tower PM. The Ulysses Tholus Caldera is 90.50° E of the Dagger Peak PM, and 62.50° W and 50.00°°° W of the Sharonov Triangle PM. Hence, the coordinates of important sites on Mars lend much support to the existence of the sacred degree.

The finding of the unnamed hexagonal-shaped crater of 1.25 degrees width at this point in my investigation is another example of blind luck. I just happened to measure one which had the correct width of a sacred degree when I was in the frame of mind to consider the possibility of the use of different degrees sizes by the Martian civilization. This is the last basic degree size that I examined seriously. Whether or not other sizes were used as well has yet to be investigated thoroughly.

The Masked Hexagon Crater

Fortuitously, there is another type of Rosetta Stone on the surface of Mars. Although it is not quite as dramatic as the Pentagon Pyramid in its

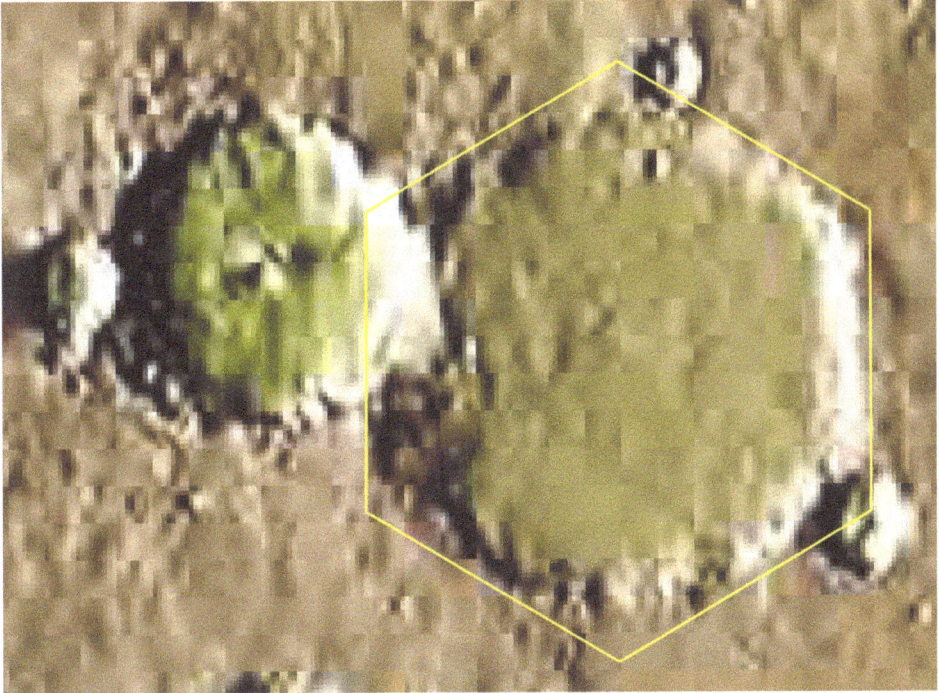

Fig. 1.9: *The Masked Hexagon Crater. A regular hexagon (yellow) was constructed to fit the outer edges of the crater perimeter. The width of the hexagon is 1.2000 latitude degrees and is 1.0066 sacred degrees of longitude. North is top of figure. USGS Astrogeology.*

confirmation of big degrees and the Dagger Peak PM, this one is more extensive since it provides evidence for the use of all 3 degree systems by the Martian architects as well as for 5 prime meridians. The site is a large unnamed crater that has the general shape of a regular hexagon (Fig. 1.9), but otherwise seems unremarkable. This 70 km wide crater is located at 139.6734° E 17.5000° S in the Terra Cimmeria region of Mars. It may be considered to be a covert hexagon since its shape is not perfectly hexagonal. Its west side is substantially intruded by another large crater making it difficult to notice the hexagonal shape. As well, its vertices are either rounded out or intruded by other craters. However, enough perimeter information is present to allow a fit to a regular hexagon which touches the edges of the crater perimeter on all sides except the southeast side which falls a bit short of the hexagon outline. Interestingly, the northeast side touches the south edge of the central peak of a small crater to the north, and the southeast side passes through what appears to be the central peak of another small crater to the southeast. These 2 craters may be considered to be auxiliary craters since they seem to be linked to the main crater for the purpose of guiding the fit of a hexagon. I sized and positioned the hexagon so as to touch the outer edges of the crater perimeter. The best fit was achieved with a hexagon whose opposite sides are 1.2000 latitude degrees apart. Because of the covert nature of the hexagon shape of the crater, I decided to call the crater the Masked Hexagon Crater.

The fitted hexagon has many remarkable properties. To begin with, despite being 1.2000 latitude degrees wide, it is 1.2583° or 1.0066°°° (sacred) longitudinal degrees in width at that latitude. It is centred at 17.5000° S in regular degrees and at 14.0000°°° S in sacred degrees. Hence, it is centred at a latitude which is an integer and 1/2 value for the 360 degree system and an integer value for the 288 sacred degree system.

Interesting longitude coordinates of the crater centre and sides are found for all 3 coordinate systems. If we use the Elysium Mons PM, we find that the centre of the crater is at 7.50° W and at 6.00°°° W. In rounded figures, the east side of the crater is at 5.50°°° W and the west side at 6.50°°° W which means that the crater is approximately 1°°° wide in longitude as mentioned above. If we use the Dagger Midline PM, the centre of the crater is at 8.50° W since this prime meridian is 1° east of the Elysium Mons PM. Less obvious though is that the east side of the crater is now at 7.00°° W (i.e., in terms of big degrees). Also, since the Sharonov Tower PM is in integer sync with both the Elysium Mons PM and the Dagger Midline PM, it too produces an integer and one-half longitude value (161.50° W) for the centre of the crater and an integer longitude value (143.00°° W) for its eastern side. Finally, using the Dagger Peak as the prime meridian, we find that the centre of the crater is at 7.49°° W. Since the Sharonov Triangle PM is in integer sync with the Dagger Peak PM for the 320 degree system, the centre of the crater is at 143.49°° W for this prime meridian. The appearance of the number 7.5 in the longitude of the centre of the crater (7.50° W [Elysium Mons PM] and 7.49°° W [Dagger Peak PM]) is very interesting in that I have found it to be the bearing angle of several very long grooves found scattered around the planet. This number would actually be 15 if alternate degree systems are used which double the number of degrees. The long grooves are discussed in detail in *Intelligent Mars III*. Other sites which we will come across in the course of this book provide additional evidence for sacred degrees (e.g., see Fig. 2.2).

So with a single crater, we have evidence for the existence of all 3 systems of degrees and the 3 prime meridians at or near Elysium Mons. Since the Sharonov Tower and Triangle also produce integer and integer and 1/2 longitudes with this hexagon, it supports those 2 candidates for prime meridians as well, bringing the total up to 5. One of the purposes of the Sharonov prime meridians seems to be to create more sacred numbers, integers, or integer and 1/2 values for the longitude coordinates of sites. Although the Sharonov Triangle PM is synchronized to the Dagger Peak PM in terms of the 360 and 320 degree systems (i.e., it is an integer number of degrees away), it is not so with the 288 degree system. This means that a site which is not synchronized to the Dagger Peak PM in terms of sacred degrees may be synchronized to the Sharonov Triangle PM and vice versa. Similar comments can be applied to the Sharonov Tower and Dagger Midline prime meridians.

There is also a heavy presence of quarter degree values for the coordinates of the hexagon. Thus the northern ends of the north-south sides of the hexagon are at 15.25°° S. Using the Dagger Peak PM, the northern and southern vertices as well as the crater centre are at 6.75°°° W, the east side is at 6.24°°° W and the west side at 7.25°°° W. This provides some evidence that quadruple degree systems may also have been used in addition to double degree systems, i.e., 4 times 320 for a 1280 degree system and 4 times 288 for a 1152 degree system.

The Masked Hexagon Crater gives us a clue as to how craters were camouflaged by the ancient Martian civilization to hide an intended geometrical shape. Enough information is given to suggest the shape but the crater is distorted to cast doubt on the correct interpretation and to leave the observer with the impression that the shape is due to natural causes rather than to intelligent engineering.

The Pavonis Mons Prime Meridians

We have seen how the longitude of the Pavonis Mons survey centre is in integer sync with the Sharonov Triangle PM being 54° W. It is also in integer sync with the Dagger Peak PM being 99° E. The longitude of the Pavonis Mons Caldera centre has even more interesting coordinates being 100° E (Elysium Mons PM), 99° E (Dagger Midline PM) and 54° W (Sharonov Tower PM). All of this suggests that the Pavonis Mons survey centre and caldera centre may be prime meridians in their own right rather than just being in integer sync with other prime meridians. The difference in longitude between the Pavonis Mons pair of sites matches the longitude difference between the dagger pair of prime meridians and the difference between the Sharonov Crater pair of prime meridians. However, in order to be plausible as prime meridians, they should show meaningful longitude differences to sites other than prime meridians. This would indicate that they were used as prime meridians in their own right to create special numbers that would not otherwise be created with the other prime meridians.

One of the most interesting longitude differences was found between

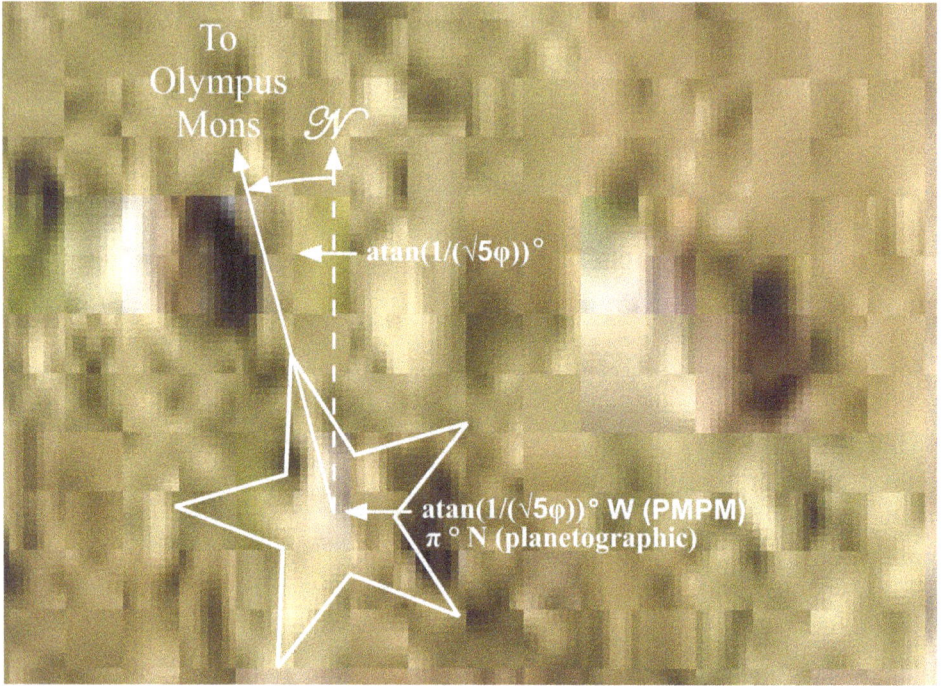

Fig. 1.10: *Symmetry of degrees of rotation of the Pentagram Pyramid from having a star point aimed due north with the longitude of the Pentagram Pyramid centre when using the Pavonis Mons survey centre as a prime meridian (PMPM). USGS Astrogeology.*

the Pentagram Pyramid and the Pavonis Mons survey centre. It was calculated to be 15.4745° which is close to atan[1/($\sqrt{5}\varphi$)] = 15.4504. This is the same value as the counterclockwise bearing angle of the northern star point of the Pentagram Pyramid which points to the survey centre of Olympus Mons, thus displaying an amazing symmetry (Fig. 1.10).

Tharsis Tholus South is 20.0177°° E and 18.0159°°° E of the Pavonis Mons Caldera, coordinates which round to 20°° E and 18°°° E. Note that 18 is half the number of regular degrees of a pentagram star point. The AscSC2* crater is 1.7300°° E of the Pavonis Mons Caldera with a numerical value very close to $\sqrt{3}$ = 1.7321. It is also 2.0139° E and 1.6111°°° E of the Pavonis Mons survey centre. The former value is close to the number 2 and the later, close to that for the golden ratio φ = 1.6180. The longitude of the Issedon Tholus survey centre is almost exactly 10φ°° E of the Pavonis Mons survey centre (16.1727 vs 16.1803). With all of this evidence, I decided that the two Pavonis Mons sites were extremely likely to have been used as prime meridians and have elected to call them the Pavonis Mons Prime Meridian (PMPM) and the Pavonis Caldera Prime Meridian (PCPM). Figure 1.11 shows where these prime meridians are located on Pavonis Mons. The total number of identified prime meridians is now 7.

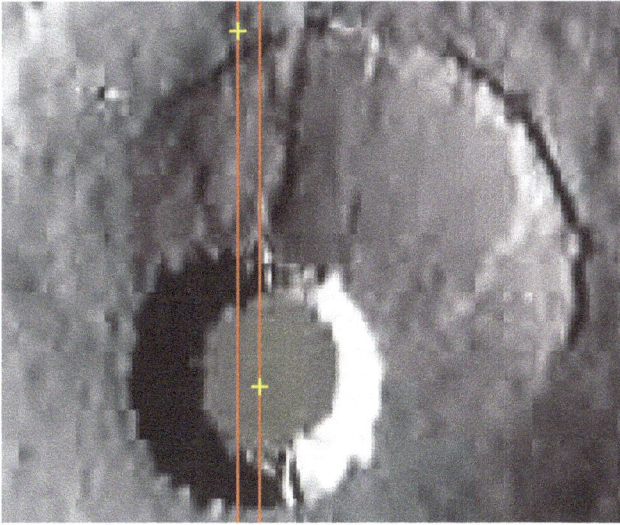

Fig. 1.11: *Location of Pavonis Mons prime meridians. The top yellow cross marks the Pavonis Mons PM at the mountain survey centre, and the yellow cross at the bottom marks the Pavonis Caldera PM. Note the rectangular area extending from the northern side of the caldera at a clockwise bearing angle of about 9 degrees. The length of the rectangle is about π times its width. USGS Astrogeology.*

Crater Edge Prime Meridian

Since 6 of the identified prime meridians occur in pairs, it seemed odd that the Elysium Mons Prime Meridian would not also have a companion prime meridian in sync with the Dagger Peak PM, the Pavonis Mons PM and the Sharonov Triangle PM. In Fig. 1.12, I positioned a meridian line on the MOLA map at the position it would occur if such a prime meridian really did exist. It aligned perfectly with the western edge of a crater lying just west of the dagger. However, to be credible as a prime meridian, it was necessary to see if it created highly meaningful longitude numbers with important sites. When used as a prime meridian, it was found that the Hecates Tholus Caldera is at 3.00° E, the HecSC1* Crater is at 5.00°° E, the north and south vertices of the Pentagon Pyramid are at 12.96°° E, Apollinaris Mons is at 27.02° E, and the centre of a crater on Albor Tholus (see Fig. 13.12) is at $\sqrt{5}$°° E. The number 3 is used extensively by the architects, often as a factor in important bearing angles. The number 5 is the number of star points in a pentagram, and the number 27 is 1/4 the size of the angle between the star points of a pentagram. The number $\sqrt{5}$ is a component of the golden mean [i.e., φ = $(\sqrt{5} + 1)/2$]. The value 12.96°° is within 2.5 min of 13°° which is a number often used on earth as the size of a group consisting of a leader with 12 followers (e.g., Jesus Christ and the 12 apostles, King Arthur and the 12 Knights of the Round Table). With these credentials, it is very possible that the western edge of the crater which is associated with the dagger by being in close proximity to it, is also intended to mark a prime meridian. Since the crater is unnamed I will simply refer to this prime meridian as the Crater Edge Prime Meridian and abbreviate it as CEPM.

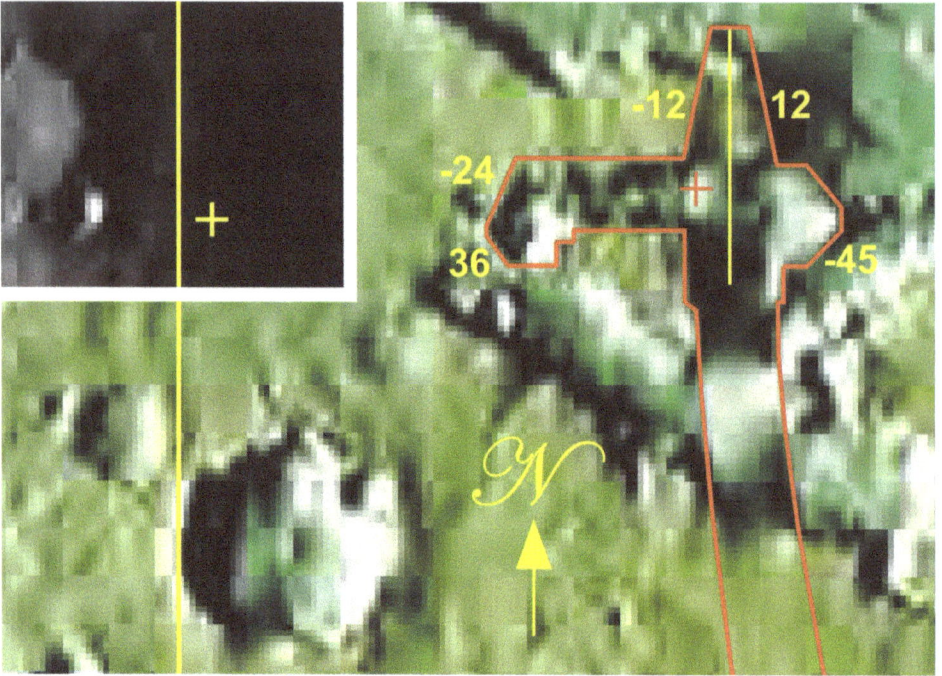

Fig. 1.12: *The Crater Edge Prime Meridian (vertical yellow line at left) lines up with the western edge of a crater just west of the dagger. Inset at top left shows the Elysium Mons survey site (yellow cross). USGS Astrogeology.*

Prime Latitudes

In *Intelligent Mars I*, it was found that the latitudes of 37 out of 48 sites were found to be either an integer or integer and 1/2 value, or could be represented by a sacred formula, or both. Thus the equator was definitely used as a prime latitude on Mars just as it is on earth. However, it was also discovered that 17 out of the 48 sites had latitude differences from the survey centre of Arsia Mons which could be expressed as sacred formulae. Of these 17 sites, 3 were also an integer or integer and 1/2 number of latitude degrees from Arsia Mons. I judged this to be strong enough evidence to declare the survey centre of Arsia Mons as a second prime latitude on Mars. We shall see that this is hugely supported by the latitudes of 5 craters on Uranius Mons (Table 13.1). All latitude coordinates in this series of books are to be considered referenced to the equator unless specifically stated to be referenced to the Arsia Mons Prime Latitude (AMPL).

The survey centre of Arsia Mons was found to be at a latitude of 8.0996° S in *Intelligent Mars I*. This value is close to $5\varphi = 8.0902°$ S. In terms of sacred degrees it is 6.4797°°° S, a value close to $4\varphi = 6.4721°°°$ S.

Conclusion

With the Pentagon Pyramid, the Masked Hexagon Crater and a crater on Hecates Tholus with a longitude of e°°° E, we have good evidence that the ancient Martians used at least 3 basic coordinate systems: the 360° system which we use on earth, the 320°° system and the 288°°° system. As noted in *Intelligent Mars I*, there is reason to believe that the Martians also used a 720° system since there were found to be a large number of integer and 1/2 longitudinal displacements from the Elysium Mons Prime Meridian. In addition we have seen the use of 1/2 degrees in sites mentioned above not only for regular degrees but also for big degrees and sacred degrees. There is even evidence with the Masked Hexagon Crater for the use of quarter degrees for the 320 and 288 degree systems, and I have noted that other sites have a longitude or latitude coordinate in quarter degrees for the 360 degree system (e.g., the north vertex of the square fitting the crater in Fig. 2.2 of Chapter 2 has a latitude of 6.25° N). A summary of the various degree systems including double and quadruple systems is presented in Table 1.2. To avoid confusion, however, I have only used the 360, 320 and 288 degree systems to express coordinate values in this book.

The 360° system is what we use on Earth today for navigational purposes and for identifying locations on the planet. Although we on Earth have anchored our 360° system to more than one prime meridian in the past, the prime meridian site was generally selected for the purpose of proclaiming military and political power. As one regime replaced

Degree System	Type	Degrees Symbol	Size (°)
360	**Regular**	°	**1.00000**
720			0.50000
1440			0.25000
320	**Big**	°°	**1.12500**
640			0.56250
1280			0.28125
288	**Sacred**	°°°	**1.25000**
576			0.62500
1152			0.31250

Table 1.2: Degree systems likely used on Mars

another, the prime meridian would shift accordingly. Since it was difficult for people to accept any change, more than one prime meridian would often be used simultaneously to satisfy the interests of several different populations. On Mars, the situation appeared to be quite different, at least in very ancient times. It would seem that the multiple degree systems and prime meridians used by the Martians were intended for much more than simply navigation and location identification. They were designed primarily to integrate important sites into Martian spirituality and metaphysics by giving them coordinates which heavily emphasized mathematical constants, integers and geometric shapes from which the Divine Creator fashioned the universe. The 288 degree system may have been created specifically for this purpose rather than for practicality since the large size of the degrees creates lower resolution. However, there was still an element of practicality because the 72 degrees of latitude between the equator and each pole could be subdivided into 9 divisions of 8 degrees each, which is superior to our own subdivision of 9 divisions of 10 degrees each due to the binary nature of the number 8.

In the rest of the book I will refer to the 8 prime meridians as: CEPM (Crater Edge Prime Meridian), EMPM (Elysium Mons Prime Meridian), DPPM (Dagger Peak Prime Meridian), DMPM (Dagger Midline Prime Meridian), PMPM (Pavonis Mons Prime Meridian), PCPM (Pavonis Caldera Prime Meridian), STrPM (Sharonov Triangle Prime Meridian) and SToPM (Sharonov Tower Prime Meridian). Since the decimal fractions of the EMPM, DMPM, PCPM and SToPM were all within 15 seconds of a degree of each other, I assumed that the prime meridians were actually integer numbers of degrees apart. I therefore used the average of their decimal part (0.1724°) when calculating coordinate positions from them. The CEPM, DPPM, PMPM and STrPM are also extremely close to an integer number of degrees apart and have an average decimal fraction of 0.1047°. The difference between the average decimal fraction for each of the 2 sets of prime meridians amounts to only 4.06 minutes of a degree. This is very close to exactly 4 minutes of a degree which may have been the intended difference. Although small, the difference between sets was very important in creating the sacred numbers necessary to convey the symbolism required by the architects. As for latitudes, it will be assumed that latitude coordinates are with reference to the equator unless specifically stated that they are with reference to the latitude of the survey centre of Arsia Mons. The Arsia Mons Prime Latitude will be abbreviated as AMPL.

The various prime meridians are summarized in Table 1.3. Also, to better orient the reader towards the location of the various prime meridians and prime latitudes, I have constructed Fig. 1.13 which shows

Table 1.3: *Summary of the 8 prime meridians likely to have been used on Mars.*

Prime Meridian	Abbreviation	Longitude (°E)		Longitude (°E)	
		Measured	Mean decimal fraction	CEPM	EMPM
Crater Edge	CEPM	147.1047	147.1047	0	
Elysium Mons	**EMPM**	**147.1701**	**147.1724**		**0**
Dagger Peak	DPPM	148.1052	148.1047	1	
Dagger Midline	**DMPM**	**148.1727**	**148.1724**		**1**
Pavonis Mons	PMPM	247.1037	247.1047	100	
Pavonis Caldera	**PCPM**	**247.1724**	**247.1724**		**100**
Sharonov Triangle	STrPM	301.1052	301.1047	154	
Sharonov Tower	**SToPM**	**301.1743**	**301.1724**		**154**

the placement of these coordinate reference lines on a picture showing about half of the planet between the latitudes of 30° S and 30° N. Since the resolution is very low at this magnification, I had to use a single line to indicate the location of the CEPM and the EMPM pair. Similarly, I had to use single lines to indicate the location of the DPPM and DMPM pair, the PMPM and PCPM pair, and the STrPM and the SToPM pair of prime meridians. The prime meridians and latitudes presented here help immeasurably to unlock the organizational plan underlying the various craters, mountains and other landforms. This tells us a lot about the civilization that designed the surface of Mars and also about our own heritage since Mars has undoubtedly played a key role in the history of planet Earth. We are now ready to explore some of the many craters on Mars to see what can be learned from these structures. In the process, the validity of the suprasystem described in this chapter will be confirmed.

* The AscSC1 (260.07° E 19.28°N), AscSC1a (264.01° E 14.01° N), AscSC2 (249.12° E 17.97° N) and HecSC1 (152.73° E 32.17° N) craters mentioned in this book were used to survey for the coordinates of mountain centres in *Intelligent Mars I*. They have no official names.

References

1. *Picture of the cover of the "Circuit" novel showing a sword marking the Prime Meridian passing through Paris.* http://www.rennes-le-chateau-la-revelation.com/dossier30.htm

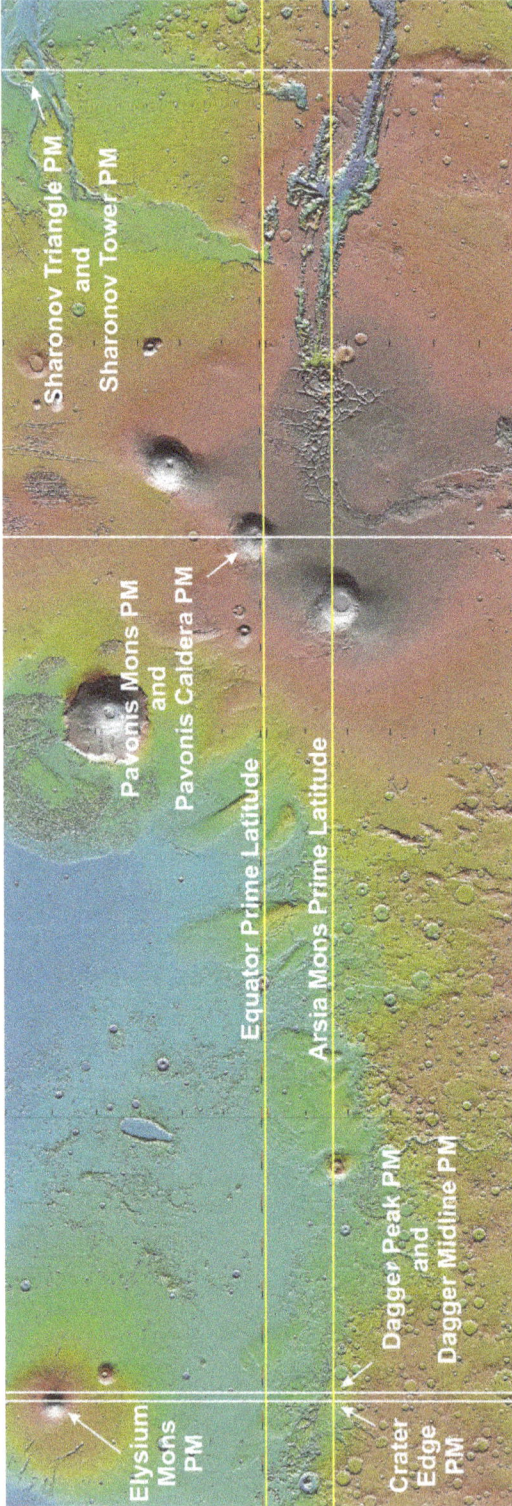

Fig. 1.13: *Placement of the 8 prime meridians and 2 prime latitudes on the planetary surface of Mars. Arrows point to the location of the markers for the specific prime meridians. Vertical white lines indicate the prime meridians. Single lines were used for the Crater Edge and Elysium Mons pair, the dagger pair, the Pavonis Mons pair, and the Sharonov Crater pair of prime meridians since the individual lines of the pairs are too close to be distinguished at this magnification. Horizontal yellow lines indicate the Equator and Arsia Mons prime latitudes. The Arsia Mons Prime Latitude passes through the survey centre for Arsia Mons at 8.0996° S. USGS Astrogeology.*

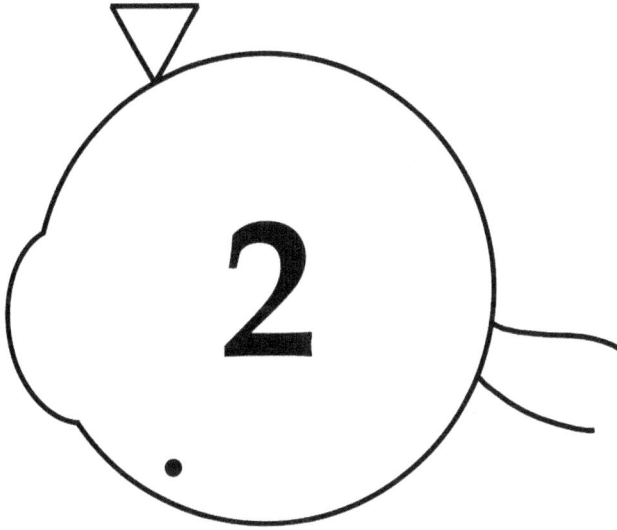

Square Craters

Although the first book of this series focused mainly on the major mountains of Mars, several important craters were also encountered, some of which functioned as survey craters, and others which played a role in Martian sacred geometry. These craters were all circular-shaped, just the shape you would expect craters to have. However, as I started to expand my search into the geometric organization of sites on Mars, the picture of perfectly circular shaped craters began to change.

Now just about everyone who looks at Mars sees only circular craters. We all grew up to consider that the good old-fashioned circular shape is the only crater shape that should occur in nature. It never registers that a substantial number of the Martian craters are not circular at all. I was guilty of this folly until I started to examine Martian craters more intensively. I began to notice that many of the craters had straight line segments in their perimeters. Then I noticed notches and steps in the perimeters of some. Other craters had such weird shapes that it required huge amounts of denial to maintain the mental illusion of their "roundness".

During my "awakening" I also started to realize that some of the craters looked almost square. Then I noticed craters shaped in the form of regular hexagons. Some even had more than 6 sides. In the next few chapters I am going to focus on craters which essentially have the shape of a regular polygon. There are many concepts fundamental to the Martian puzzle to be learned from them. A regular polygon is a 2 dimensional geometric shape which has straight line sides which are all equal in length and whose internal angles between adjacent sides are all equal in size. Besides the square and the regular hexagon, I have also discovered craters which are regular pentagons and regular octagons. These crater shapes are rarely fully true regular polygons since corners are often rounded and perhaps one of the sides might be ill-fitted. But enough perimeter information exists firstly to unmistakably identify the crater to be at least closely mimicking one having a definite number of equal sides and angles, and secondly to permit a reasonably accurate fit of a regular polygon to its perimeter outline. Many other polygonal-shaped craters also exist on Mars which are not regular. These will not be covered in this book.

Since non-circular craters are not natural, it should be obvious that many of the craters on Mars could not have been caused by meteor impacts. They could only have been artificially constructed. The material that I am now going to present will only confirm that conclusion. This chapter will focus on square-shaped craters. The other polygonal-shaped craters will be presented in subsequent chapters.

Introduction to Square Craters

I will start with one of the smaller "square" craters. It is an unnamed crater which I identified as the nn5 crater in my own numbering system. It has 4 long straight line segments which are at right angles to each other (Fig. 2.1 top left). The straight segments on the southeast and southwest sides are intermittent whereas on the northeast and northwest sides they are more continuous. There is also a 5th straight line segment which travels in a north-south direction on the crater's west side which is not as long as the other 4. This crater is well fit by a square with its centre located at 61.7861° E 9.1393° S and having a diagonal size equal to 0.6667 latitude degrees (Fig. 2.1 top right). The square is rotated so that its northwest side has a bearing angle of 45 degrees in the clockwise direction. Thus its 4 vertices point exactly to the 4 cardinal points of the compass with the northern vertex at 0.7064° S with reference to the Arsia Mons Prime Latitude. This number is very close to $\sqrt{2}/2 = 0.7071$. In terms of the ancient Martian prime meridians, the centre of the square is located at 86.3863° W of the Dagger Midline PM with a numerical value close to $33\varphi^2 = 86.3951$. Its eastern vertex is 212.4275°° W of the Sharonov Triangle PM with a numerical value very close to $95\sqrt{5} = 212.4265$. Keeping the same centre, if we rotate the fitted square either clockwise or

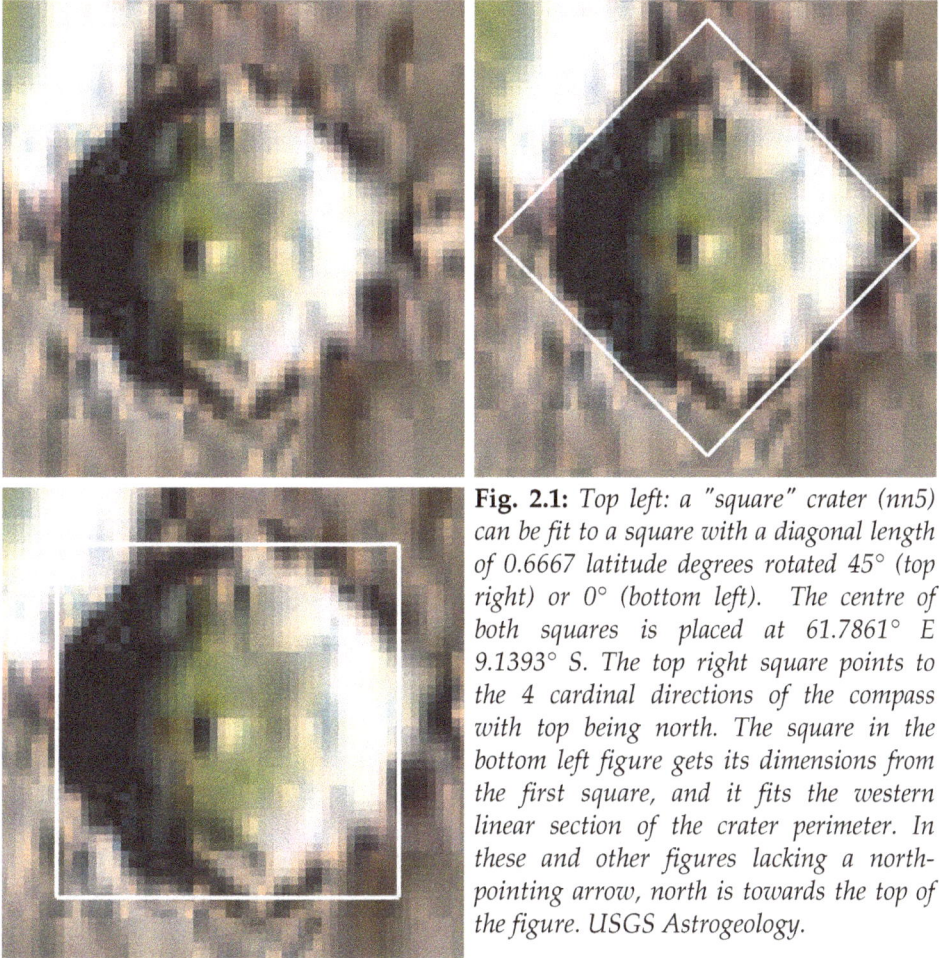

Fig. 2.1: *Top left: a "square" crater (nn5) can be fit to a square with a diagonal length of 0.6667 latitude degrees rotated 45° (top right) or 0° (bottom left). The centre of both squares is placed at 61.7861° E 9.1393° S. The top right square points to the 4 cardinal directions of the compass with top being north. The square in the bottom left figure gets its dimensions from the first square, and it fits the western linear section of the crater perimeter. In these and other figures lacking a north-pointing arrow, north is towards the top of the figure. USGS Astrogeology.*

counterclockwise by 45 degrees so that 2 of its sides are parallel to lines of latitude, we find that the square now fits the west straight line segment (Fig. 2.1 bottom left). The eastern corner of the crater perimeter, however, extends well beyond the eastern side of the square. The north and south sides of the square fit small segments of the crater perimeter but short sections of the perimeter extend beyond the fitted square. The southern side of this second square is at 7.5000°°° S. The west side of the square is 77.0000°° W of the Dagger Midline PM and 165.0000°° W of the Pavonis Caldera PM. It is also 76.1111°° W of the Elysium Mons PM and 76.0510°° W of the Crater Edge PM. These numbers are close to 28e = 76.1119 and 47φ = 76.0476. The east side is at 68.0639°°° W of the Crater Edge PM with a numerical value close to $26φ^2 = 68.0689$. This crater demonstrates that "square" craters can sometimes be fit by 2 or more squares differing only in the bearing angles of their sides.

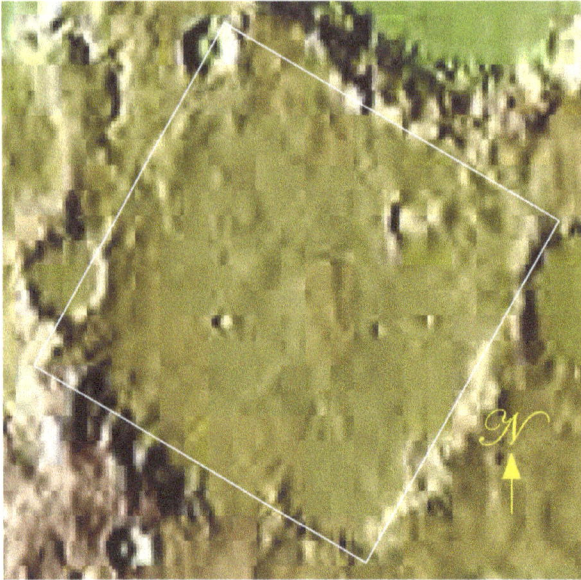

Fig. 2.2: *A square with a diagonal length of 2 latitude degrees whose northwest side has a bearing angle of 30° clockwise can be nicely fit to this unnamed crater (nn1) at 39.7349° E 5.2842° N. The northern vertex is at 6.2501° N and 5.0001°°° N. The western vertex is at 10.5000°°° N (AMPL). USGS Astrogeology.*

The second "square" crater which I will show you is a very shallow but large crater (Fig. 2.2). The crater is also unnamed and is identified as the nn1 crater in my numbering system. This crater can be best fit with a square centred at 39.7349° E 5.2842° N and having a diagonal length of exactly 2 latitude degrees (118.5 km). Its northwest side has a bearing angle of 30° in the clockwise direction. This crater illustrates that parts of a "square" crater can be intruded by other structures which tend to mask the full perimeter of the fitted square. Thus the northeast corner of the fitted square lies just inside the walls of a smaller crater to the east. Its northwest side passes through 2 other smaller craters. But notice that this side of the square passes through the middle of the central peak of the northern small crater and that there are interruptions in the south and north perimeters of the southern small crater which align to the northwest side of the square. The rest of the fitted square is well delineated by very long linear stretches of the crater perimeter although the square aligns mainly to the base of the crater wall rather than to its outside edge for much of its course. The northern vertex of the square is at 6.2501° N and 5.0001°°° N and the centre of the square is at 95.5000°° W of the Elysium Mons PM. The western vertex of the square is 10.5000°°° north of the Arsia Mons Prime Latitude (AMPL).

The Teisserenc de Bort Crater is another crater that lends itself to fitting with a square. It is a very large crater whose centre sits almost on the equator, and gets its name from the French meteorologist who is credited with the discovery of the stratosphere. This crater is best fit by a square whose diameter is equal to 2.6667 latitude degrees and whose centre is located at 45.0172° E 0.4642° N (Fig. 2.3a). The square is rotated so that the bearing angle of its northwest side is 30 degrees in the clockwise direction. The northeast and southwest sides of the square run along lengthy linear sections of the crater perimeter. On the northwest side, the square runs along the edges of 2 small medium brown regions (2 upper arrows, left), touches the edge of a notch in the crater perimeter (middle arrow) and also runs along the northwest linear edge of a smaller crater to the southwest of the Teisserenc de Bort Crater (lower arrow). The southeast side of the square does not have any noticeable alignments. The north vertex of the square lies at 82.0003°°° W of the Elysium Mons PM, 92.0004°° W of the Dagger Midline PM, a very notable 180.0004°° W of the Pavonis Caldera PM and 228.0004°° W of the Sharonov Tower PM. The latitude of the northern vertex is also remarkable in that it is close to $\sqrt{3}$° N (1.7521° vs. 1.7321°), $\pi/2$°° N (1.5574°° vs. 1.5708°°) and $\sqrt{2}$°°° N (1.4017°°° vs. 1.4142°°°). The west vertex of the square is 83.5003°°° W of the Dagger Peak PM. The east vertex is very close to $\varphi/2$° N (0.8093° vs. 0.8090°). Thus this remarkable square honours all the most basic irrational numbers except $\sqrt{5}$ and e in the latitude coordinates of its vertices.

If the square is reduced in size so that its diagonal is now equal to 1.875 latitude degrees, and the centre remains the same as the larger square (Fig. 2.3b), the southwest side runs along a linear section of the

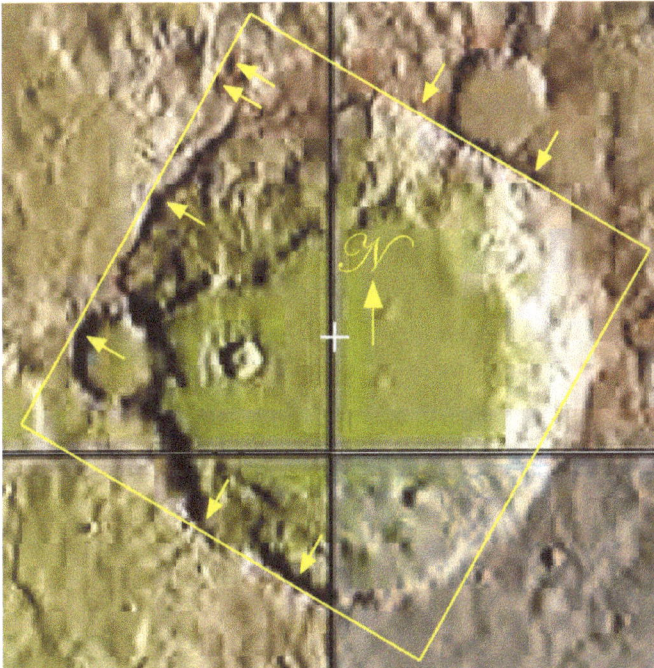

Fig. 2.3a: *The Teisserenc de Bort Crater can be well fit to a square whose diagonal is equal to 2.67 latitude degrees and whose northwest side has a bearing angle of 30° in the clockwise direction. Yellow arrows mark where the fitted square aligns with the crater perimeter and other structures. A white cross marks the centre of the square. USGS Astrogeology.*

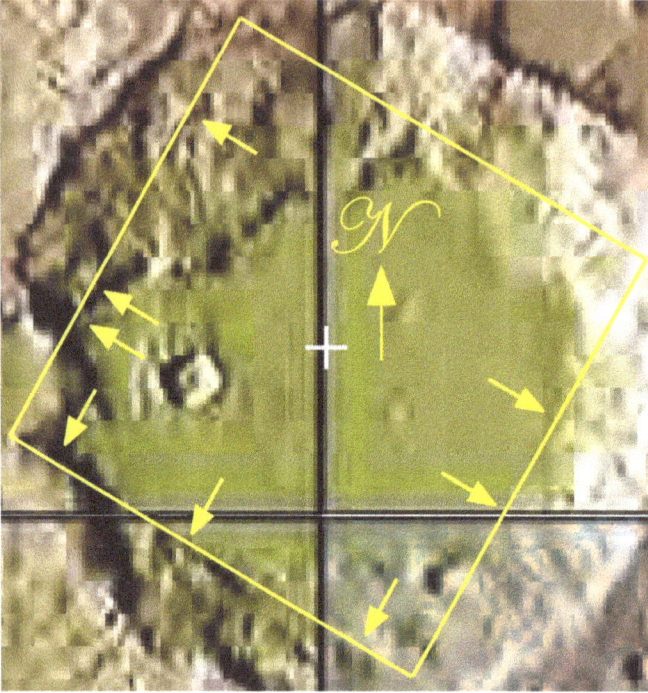

Fig. 2.3b: *If the diagonal size of the square fitting the Teisserenc de Bort Crater is reduced to 1.875 latitude degrees, the square aligns to other elements of the crater (yellow arrows). This is an example of a crater that can be fit to more than one size of square. Black lines are map gridlines. A white cross marks the centre of the square. USGS Astrogeology.*

crater perimeter (lower left arrow) as well as along a lengthy part of the inner border of the crater wall (middle arrow at bottom). Close to the south vertex, the southwest side runs along the junction of dark and light areas (bottom arrow). The northwest side aligns to the bottom edge of a short section of the crater wall (lowest arrow pointing west). It also aligns to 2 regions where there is a transition between dark and light coloured areas (the middle and top west pointing arrows). The southeast side appears to align with the lower edge of the crater wall in a couple of locations (yellow arrows) although the border is not well demarcated on the map. The northeast side of the square does not seem to align to any feature of the crater. The east vertex of the smaller square has the remarkable longitude of 89.9996°° W and 80.9997°°° W of the Elysium Mons PM since the numbers round to 90 and 81 respectively. The western vertex is at 103.9930° W of the Dagger Peak PM, and 256.9930° W of the Sharonov Triangle PM. These numbers round to 104 and 257 degrees respectively. The latitude of its northern vertex is close to the value for e/2° N (1.3698° vs. 1.3591°), and the latitude for the east vertex is close to $\sqrt{2}/2$° N (0.7068° vs. 0.7071°) or $\sqrt{5}/4$°°° N (0.5654°°° vs. 0.5590°°°). Hence, this square honours the basic irrational numbers that were missed by the larger square and, as well, repeats a reference to $\sqrt{2}$.

The Teisserenc de Bort Crater is an example of a "square" crater which can be fit to more than 1 size of square, whereas Fig. 2.1 showed an example of a "square" crater that can be fit to more than 1 rotation of the same size of square. The craters in Figs. 2.1 - 2.3 also demonstrate how portions of a fitted square can extend beyond the boundaries of the crater. These are some of the basic concepts about "square" craters. I will now proceed to describe some other very interesting "square" craters before coming to an overall assessment of this class of crater.

The Nicholson Crater

The Nicholson Crater (Fig. 2.4) is a good example of a covert "square" crater, a crater that doesn't look particularly square but can nevertheless be well fit by a square. The Nicholson Crater, named after the American astronomer Seth Nicholson, has the appearance of a round crater to the casual observer. Close examination, however, reveals several straight line segments in its perimeter which are parallel to one another on opposite sides of the crater. Some of these can be well fit to a square whose diagonal is 2.5 latitude degrees and whose northwest side has a bearing angle of 45° in the clockwise direction (Fig. 2.5). The northwest, northeast and southeast sides fit very well to straight line segments of the crater perimeter, but the southwest side lies well south of the straight line segment of the perimeter. The southwest side nevertheless touches the

Fig. 2.4: *The Nicholson Crater appears to be round to the casual observer. The yellow arrow points to the approximate location of the central peak of the crater which is 3.5 km higher than the crater floor. The black line marks the equator. USGS Astrogeology.*

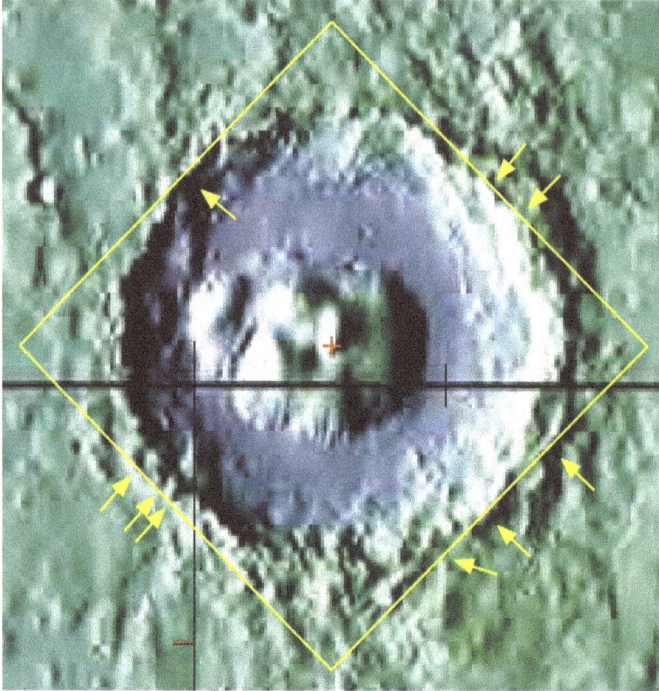

Fig. 2.5: *A square with a diagonal size of 2.5 latitude degrees and a clockwise bearing angle of 45° for its northwest side is nicely fit to several straight line segments in the Nicholson Crater perimeter. Yellow arrows mark these and other points of alignment to the square. The red cross marks the centre of the square and is located at 195.5474° E 0.1504° N. The horizontal black line is the equator. USGS Astrogeology.*

tips of 3 fingers (pointed out by 3 arrows) which protrude out from the crater and thus actually bounds the crater perimeter properly on this side. The southeast side of the square also aligns to a light coloured structure (bottom arrow) and to a light coloured straight line region (upper arrow) lying outside the crater perimeter. The coordinates of this square are very interesting. The centre of the square together with the east and west vertices are at the latitude of 0.1337°° N whose value is close to $1/e^2$ = 0.1353. The northern vertex is at a latitude of 1.1203°°° N which is close to the value for $\sqrt{5}/2$ = 1.1180. It is also at 9.500° N (AMPL). The latitude for the southern vertex is 0.8797°°° S which is close to the value for $\sqrt{2}/\varphi$ = 0.8740. Not so obvious though is that this latitude is at exactly 7.0000° N (AMPL). The longitude of the centre of the square and that of its north and south vertices is equal to 43.0000°° E of the Elysium Mons PM. The longitude of its western vertex is 41.0000°° E of the Dagger Midline PM, 47.0000°° W of the Pavonis Caldera PM and 95.0000°° W of the Sharonov Tower PM. In terms of sacred degrees, the east vertex is at 83.5000°°° W, the north and south vertices and centre of the square are at 84.5000°°° W and the west vertex is at 85.5000°°° W, all in reference to the Sharonov Tower PM. Since they are integer and 1/2 degrees, these numbers are consistent with the theory that a 576 degree system was in use on Mars as well as a 288 degree system. It also means that the diagonal of the square

Fig. 2.6: *If the square fitting the Nicholson Crater is reduced to a diagonal size of 2.25 latitude degrees and the centre stays at the same location, the southwest side fits the crater perimeter very well. Yellow and red arrows mark other points of alignment to the square. Black line marks the equator. USGS Astrogeology.*

parallel to the equator is exactly equal to 2 sacred degrees. This works out since the Nicholson Crater is extremely close to the equator where a degree of longitude is equal in distance to a degree of latitude. Thus 2.5 latitude degrees divided by the size of a sacred degree (1.25 standard degrees) works out to exactly 2 sacred degrees. Also of interest is that the centre of the square (red coloured cross) comes very close to and may be coincident with the location of the central peak of the crater which is 3.5 km higher than the crater floor. The central peak has the appearance of a pyramid when viewed from the south since its southern side is flat, whereas the side on the northern half is rounded. Beautiful imagery from the High Resolution Stereo Camera (HRSC) of the Nicholson Crater and its central peak can be seen at http://www.youtube.com/watch?v=9bXIqq8RcSg.

If the size of the square is reduced from a diagonal size of 2.5 to 2.25 degrees of latitude and the location of the centre is maintained, a remarkable thing happens. The southwest side now exactly lines up with the long linear segment of the crater perimeter (Fig. 2.6). The northwest and northeast sides line up with short linear segments of the crater perimeter, and the southeast side lines up with the northwest edge of a structure lying beyond the crater perimeter (see red arrow). Since this new square has the same centre and bearing angle as the larger one, it

Fig. 2.7: *Large square of Fig. 2.5 rotated 15° counterclockwise (yellow square) and 15° clockwise (white square) to give clockwise bearing angles of 30° and 60° respectively to their northwest sides. Alignments with the Nicholson Crater are shown by short red lines superimposed on the squares. See text for further explanation. USGS Astrogeology.*

shares several of its coordinate values. However, the longitudes of its east and west vertices and the latitudes of the north and south vertices are changed. The west vertex is at 42.0000°° E of the Elysium Mons PM and 37.0000°°° E of the Dagger Midline PM. Its east vertex is at 49.5000° E and 44.0000°° E of the Elysium Mons PM, 48.5000° E of the Dagger Midline PM, 50.5000° W of the Pavonis Caldera PM and 104.5000° W of the Sharonov Tower PM. The southern vertex has a latitude of 0.8663°° S which is very close to $\sqrt{3}/2 = 0.8660$, and the northern vertex is 7.5000°°° N of the Arsia Mons Prime Latitude (AMPL).

Now let's rotate the large square (diagonal of 2.5 degrees) counterclockwise by 15° so that the bearing angle of its northwest side is 30° instead of 45° (yellow square of Fig. 2.7). This square lines up with 2 short linear regions of the crater perimeter on the northeast side (red lines superimposed on square). The southeast side aligns with the edge of the same structure lying beyond the crater perimeter as in Fig. 2.6 but at a slightly more southerly location. The square's northwest side aligns with a long section of the outside edge of the crater apron instead of its perimeter. The southwest side aligns with the south edge of a small crater near the west vertex of the square and with the edge of a structure beyond the crater perimeter. If the original large square is rotated by 15° in the clockwise direction so that the bearing angle of the northwest side is now

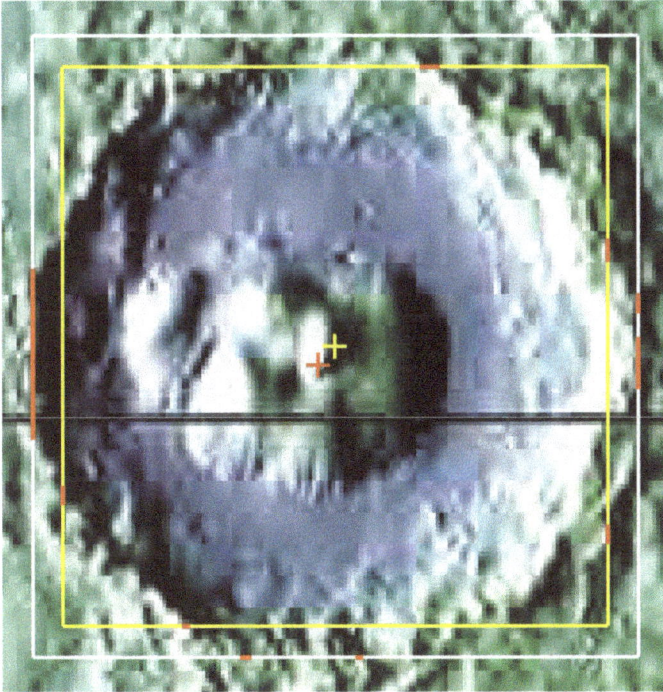

Fig. 2.8: *Two sizes of square fitting the Nicholson Crater which have diagonal sizes of 2.5 and 2.25 lateral degrees, and have their west sides at a bearing angle of 0°. These squares align to linear regions of the crater perimeter and other structures (short red lines). The centre of these squares, shown as a yellow cross, differs from the centre of the other previously shown rotations (red cross). It is located at 195.5958° E 0.2045° N. USGS Astrogeology.*

60° we obtain the white square seen in Fig. 2.7. This square fits a long linear segment of the Nicholson Crater perimeter on the northeast side, and short segments on the northwest and southeast sides. The southeast side also aligns to a couple of structures lying beyond the crater perimeter. The southwest side aligns to the end of a small finger projecting out from the crater interior. The northern vertices of the 2 squares are at 1.3578° N which is close to 1.3591 = e/2° N. The western vertex of the yellow square and the eastern vertex of the white square are at 0.1731° S which is very close to 0.1732 = √3/10° S. The longitude from the Dagger Peak PM of the northern vertex for the yellow square and for the southern vertex of the white square is 47.1191° E, 41.8836°° E and 37.6953°°° E. The amazing thing is that these numbers are all very meaningful since they are close to 15π = 47.1239, 16φ² = 41.8885 and 12π = 37.6991.

There is yet one more rotation to consider. If the original large square is rotated by 45 degrees, the north and south sides of the square become parallel to lines of latitude, and the east and west sides line up with linear sections of the crater perimeter (see short red lines on white square of Fig. 2.8). But in order to do this, the centre of the square had to be shifted eastwards by 0.0484°. I also shifted it northwards by 0.0541° (see below). The east side of the square sits at 43.0000°° E (Dagger Peak PM) and 93.0000°° W (Sharonov Triangle PM). If this square is now reduced to a

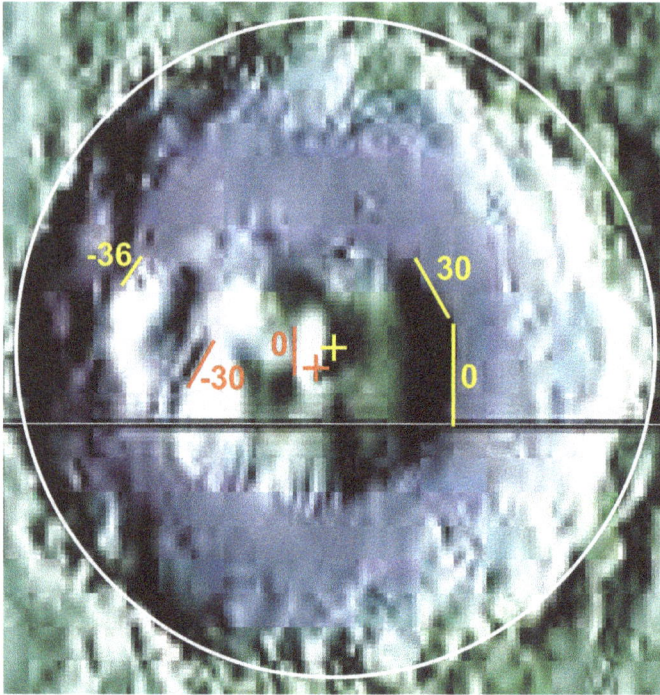

Fig. 2.9: *Straight line segments and their bearing angles for different features of the massive structure inside the Nicholson Crater. These lines and angles suggest artificiality. The white circle fitting the perimeter is centred at the yellow cross which is the same centre used for the squares in the previous figure. USGS Astrogeology.*

diagonal size of 2.25 latitude degrees and the centre placed at the same new coordinates, the yellow square of Fig. 2.8 is obtained. This square lines up with small steps in the crater perimeter for both the east and west sides. The east side also aligns with the edge of a structure lying beyond the crater perimeter. The longitude of the west side of the smaller square is 41.5071°° E (Dagger Peak PM) and 94.4929°° W (Sharonov Triangle PM). These numbers round to 41.5 and 94.5. I was unable to ascertain the latitude of the centres of the 2 squares with any great certainty. I simply placed their centres (yellow coloured cross) at the latitude for which the northern side of the smaller square was exactly at 1.0000° N. This resulted in its southern side latitude to be at 6.0069°°° N (Arsia Mons PL). This placement also produced some minor alignments to the Nicholson Crater on the south sides of the 2 squares and an alignment to a small step in the crater perimeter for the north side of the smaller square.

With the fitting of the various squares, we have ample evidence of the artificiality of this crater. This interpretation is further enhanced when you look at the massive structure in the crater interior. There are several straight line edges in the structure which have bearing angles of 0°, 30° and 36° (Fig. 2.9). Being about 55 km long by 37 km wide, this structure is large enough to house an entire city. It was also found that a circle could be well fit to the crater perimeter with its centre at the yellow cross

Fig. 2.10: *The Henry Crater located at 23.4774° E 10.7500° N. The crater has a square-like appearance with rounded corners. The structure inside also has a square-like appearance. USGS Astrogeology.*

corresponding to the centre of the 2 squares just discussed above. This suggests that my choice of latitude for these 2 squares was correct after all. It also demonstrates that the eye was correct in assessing that the crater has an essentially round shape. This underlying shape was cleverly distorted with relatively small deviations in appropriate regions to permit the fitting of squares.

The Henry Crater

The Henry Crater (Fig. 2.10) is named after the 19th century French telescope makers and astronomers Paul-Pierre Henry and Mathieu-Prosper Henry (brothers). It is a huge crater about 170 km in diameter with an enormous structure in its interior. The crater has a square-like

Fig. 2.11: *The Henry Crater can be well fit with a square having a diagonal length of 4 latitude degrees and clockwise bearing angle of 45° for its northwest side. The square aligns to linear sections of the crater perimeter on all sides except on the southwest side. The southeast side aligns to the edges of 2 small craters (red arrows) as well as to the crater perimeter. USGS Astrogeology.*

appearance with rounded corners. The interior structure itself is almost square with a side length of about 90 km. Like the structure in the Nicholson Crater, it could contain a large city within its dome. After considerable experimentation, I decided that a good way to fit the crater perimeter was with a square with a diagonal size of 4 latitude degrees whose northwest side has a bearing angle of 45° in the clockwise direction (Fig. 2.11). The size of 4 latitude degrees is an interesting way to refer to the 4 sides of a square. Also, the side length of this square is $2\sqrt{2}$ latitude degrees. The value of $\sqrt{2}$ is another reference to the square since it is a factor in the length of the diagonal of a square.

Besides its integer diagonal size, this square has some special properties. It fits the outermost linear regions of the perimeter on 3 sides (see short red lines superimposed on the white square in Fig. 2.11). On the southeast side, the square aligns with the edges of 2 small craters (red arrows). The northwest side also aligns with a linear trench just inside the base of the perimeter (upper red line on this side). The southwest side fails to line up with any feature of the crater or other landmarks. The centre of the square, marked with a white cross, is located at 23.4774° E 10.7500° N. Its latitude becomes interesting when referenced to the Arsia Mons Prime Latitude in which case it is calculated to be 18.8496° N (AMPL). The number 18.8496 is equal to 6π. When referenced to Martian prime meridians, the centre of the square has a longitude of 110.8400°° W (Dagger Midline PM) and 109.9511°° W (Elysium Mons PM). These longitudes seem like uninteresting numbers until you realize that $15e^2$ = 110.8358, 35π = 109.9557 and $42\varphi^2$ = 109.9574. The longitude of the western vertex also has interesting numbers. It is 100.5845°°° W of the Elysium Mons PM, 112.5893°° W of the Dagger Peak PM, and 125.6630° W and 100.5304°°° W of the Crater Edge PM. These numbers closely translate into 37e (100.5764), $43\varphi^2$ (112.5755) and $65\sqrt{3}$ (112.5833), 40π (125.6637) and $48\varphi^2$ (125.6656), and 32π (100.5310). The eastern vertex has a longitude of 177.3274°°° W of the Pavonis Caldera PM where $24e^2$ = 177.3373. The southern vertex has a much simpler latitude of 7.0000°°° N. The northern vertex has a latitude of 11.3333°° N which creates an interesting symmetry with the southern vertex since 7φ is equal to 11.3262. Also, if you can remember, the latitude of the southern vertex of the first square that I used to fit the Nicholson Crater (Fig. 2.5) is 7.0000° N but with reference to the Arsia Mons Prime Latitude. In addition, I found the use of 7 (see Table 2.3 below) for the latitudes of the vertices of 3 more square-shaped craters.

After completing my study of the fit of this first square to the Henry Crater, I turned my attention to the long, relatively linear, section of the crater perimeter on the west side which was in a north-south orientation. It occurred to me that if I simply rotated the square by 45 degrees in the counterclockwise direction about its centre, the new orientation might fit some of the western crater perimeter since the sides of the square would now be approximately parallel to lines of longitude. The new square is shown in Fig. 2.12 (largest square) and its western side does indeed line up with portions of the western crater perimeter as indicated by 3 short red lines superimposed on the square. This side of the square also passes directly through the middle of the central peak of a fairly large crater adjacent to the southwest corner of the Henry Crater. The eastern side of

Fig. 2.12: *The outer square is the same square from Fig. 2.11 rotated 45 degrees counterclockwise without changing the location of the centre. The middle square is 5/6 the size of the outer square. The inner square is 3/5 the size of the outer square. Short red lines superimposed on the squares denote alignments of the squares with straight line edges associated with the Henry Crater and auxiliary craters. USGS Astrogeology.*

the square aligns to a short section of the Henry Crater eastern perimeter and passes through the centre of a small crater near the southeast vertex. The south side of the square aligns to the edge of a small pocket of low terrain on the south side of the Henry Crater perimeter. There does not appear to be any substantial alignments for the north side of the square. The north side of the square has a latitude of 18.0123°° N (AMPL) and the south side a latitude of 15.4981°° N (AMPL). These values are less than 1 min of a degree from 18°° N (AMPL) and 15.5°° N (AMPL). The number 18 is half the number of degrees in a star point of a pentagram, a geometric figure which plays a very important role in the sacred geometry of the Martian mountains as discussed in *Intelligent Mars I*. The east side of the square has an average longitude of 221.0040°°° W (Sharonov Tower PM) which has a numerical value close to 221. Note that since I constructed all dimensions of the square in terms of latitude degrees, the longitudes of the east and west sides are not a constant, particularly since this square is very large. Lines of longitude are closer together at larger latitude values than at smaller latitude values.

I next reduced the size of the latest square (keeping the centre constant) until its southern side fit the southern edge of a small crater lying just east of the big crater at the southwest corner of the Henry Crater (see middle square in Fig. 2.12). This also aligned with a step in the eastern perimeter of the big crater. The size of the new square is exactly 5/6 that of the original square, giving it a diagonal size of 3.33 latitude degrees. The west side of the square aligned with a step in the northeast perimeter of the southwest crater and also with the eastern edge of a light coloured structure within the Henry Crater further to the north. The east side of the square passed directly through the middle of the central peak of a small crater lying just east of the Henry Crater's southeast side. There were no notable alignments with the north side of the square. What drew my attention to this square is that the latitude of the southern side is 14.1369°°° N (AMPL). The value of 14.1421 is equal to $10\sqrt{2}$. The involvement of the $\sqrt{2}$ value is noteworthy since the size of the diagonal of a square is $\sqrt{2}$ times the length of one of its sides. The latitude of this side also honours other irrational numbers since it can be expressed as 15.7076°° N (AMPL) where the value of 15.7082 is equal to $6\varphi^2$ and 15.7080 is equal to 5π. These latter numbers might be a reference to the pentagram since that geometric figure contains several different measures of the golden ratio $\varphi = 1.6180$ and has 5 star points. The average longitude of the east side of the square is 97.9961°°° W (Elysium Mons PM) and 177.9961°°° W (Pavonis Caldera PM), numbers which round to 98 and 178.

There is one more square which I will discuss for the Henry Crater. If the largest square in Fig. 2.12 is reduced to 3/5 of its original size giving it a diagonal size of 2.40 latitude degrees, a square is produced which exactly lines up with the west linear side of the perimeter of a crater inside the eastern part of the Henry Crater (see smallest square in Fig. 2.12). No other alignments could be seen for this square. Its south side has a latitude of 9.9015° N which has a numerical value very close to $7\sqrt{2}$ = 9.8995. Its latitude is also 18.0011° N (AMPL) and 16.0010°° N (AMPL). The number $\sqrt{2}$ could refer to the diagonal of a square, the number 18 is half the number of degrees in the star point of a pentagram, and 16 = 4^2 could refer to the number of sides in a square. Note that 18 is in terms of regular degrees for the south side of this square whereas it was in terms of big degrees for the north side of the largest square above. The average longitude of the west side is 100.3929°°° W of the Dagger Peak PM where the numerical value is close to 71√2 = 100.4092. Remarkably, the average longitude of the east side is 99.0108°°° W (DPPM) where the numerical value is close to 70√2 = 98.9949. This means that the length of the side of the square must be very close to √2 sacred degrees of longitude at this latitude. It calculates out to be an average of 1.3821°°° rather than 1.4142°°° so it is actually about 2.3 % smaller than √2. Nevertheless, the intent may have been to refer to √2.

The Henry Crater is one more example of how a "square" crater can be fit to more than 1 rotation of square and to more than 1 size of square. It is also an example of how smaller craters both outside and internal to the larger "square" crater are used for alignment purposes with their edges, centres or peaks. I decided to call these guiding craters *auxiliary* craters as I did for craters associated with the Masked Hexagon Crater.

Elysium Mons Caldera and Albor Tholus Caldera

The Elysium Mons Caldera is a special example of a "square" crater. Although it is a caldera rather than a crater, it mimics the "square" craters and has all the earmarks of artificiality. To start with, it is square-shaped rather than round (Fig. 2.13). It is oriented so that the 4 vertices of the square are pointing in the cardinal directions of the compass, which is very appropriate since the triple grouping of Elysium Mons, Albor

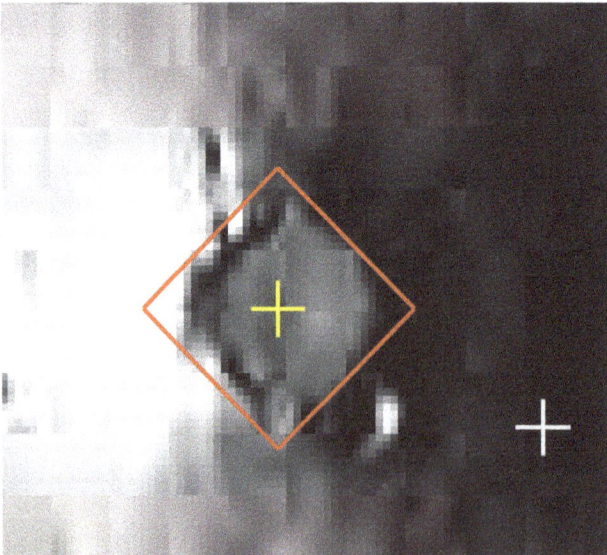

Fig. 2.13: *The caldera of Elysium Mons is well fit by a square whose diagonal size is 0.375 latitude degrees and whose northwest side has a bearing angle of 45°. The yellow cross is at the centre of the caldera square and the white cross is at the survey centre of Elysium Mons. USGS Astrogeology.*

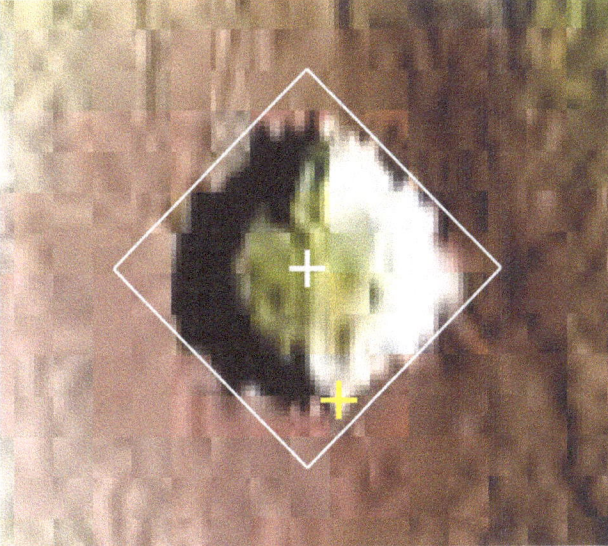

Fig. 2.14: *The caldera of Albor Tholus is well fit by a square whose diagonal size is 0.8333 latitude degrees and whose northwest side has a bearing angle of 45°. The white cross is at the centre of the caldera square and the yellow cross is at the survey centre of Albor Tholus. USGS Astrogeology.*

Tholus and Hecates Tholus serve as a compass with the 2 tholi lying on the same longitude line thus pointing to the north and south poles. The diagonal size of the square which fits the outermost edges of the perimeter of the Elysium Caldera is 0.375 latitude degrees. It shows a good fit when the east vertex of the square is positioned at 1.0000°° W of the Dagger Peak PM and the south vertex at 9e = 24.4645° N and 8e = 21.7463°° N! Note that 9 is 1/4 the size of the angle of the star points of a pentagram and that 8 is a Fibonacci number which relates to φ. Note also that 8 is twice the value of 4, the number of equal sides in a square. For all of this to happen by chance is unimaginable. It is another piece of evidence that the mountains themselves have been artificially constructed at predetermined locations and are not naturally volcanic in origin.

The caldera on top of Albor Tholus is also "square" with its 4 vertices pointing in the cardinal compass directions. A square with a diagonal size of 0.8333 latitude degrees gives a good fit when it is centred at 2.7324°° E of the Elysium Mons PM which is close to the value for e = 2.7183. The centre is also exactly at 3.1416 = π° E of the Crater Edge PM and 1.7133°°° E of the Dagger Peak PM which is close to the value for √3 = 1.7321. The latitude of the northern vertex of the square is very remarkable and is analogous to the latitude of the south vertex of the Elysium Mons square. It is located at 10√3 = 17.3205°° N and 9√3 = 15.5885°°° N, giving us even more evidence for artificiality. Thus although the Albor Tholus Caldera refers to e, it is more focused on π and √3 whereas the Elysium Mons Caldera is focused only on the value for e. Note the parallel in the use of the number 9 in the coordinates of the squares for both caldera.

Decoding Square Craters

With the above examination of several "square" craters, you can see that a picture is starting to form as to their general properties. Square craters come in many different sizes. They often provide a template for more than 1 square in which the replicates can differ in both size and bearing angle. Some are clearly very square but more often their square nature is rather covert. Sometimes lengthy linear sections of a crater's perimeter are present to demarcate the location of the sides of a square. Other times only a short linear length is available for alignment of a side and occasionally a side is not aligned to any feature of the crater. Sometimes alignments occur with auxiliary craters or with structures inside or outside the main crater. Corners of "square" craters are most often quite rounded which helps to camouflage them.

Is there anything else that they can reveal? What about their size? Do some sizes occur more often than others, and is there a pattern as to size? Do they form some sort of grid to mark out the positions of lines of latitude and longitude in a systematic way? To try to answer these questions, I analyzed all the craters that I could readily identify as being "square" between 30° S and 30° N which were the latitude limits of the USGS Mercator MOLA maps that I was using. I found a total of 26 "square" craters of which 7, including the 2 "square" calderas, have already been presented. For most of these craters, I limited myself to the square which fit the outermost reaches of the crater perimeter and had the bearing angle which fit the longest linear segments of the perimeter for that size of square. For the nn5 crater, the Teisserenc de Bort Crater, the Nicholson Crater and the Henry Crater, I also fit more than 1 rotation and/or size as shown above.

The data from this exercise are presented in Table 2.1 in which squares fitting the "square" craters are listed in order of diagonal size in latitude degrees. Many of the "square" craters do not yet have official names so I have identified them with my own numbering systems with which I assigned to craters during the course of my studies over several years. For craters with more than 1 size or rotation of square, I appended the letters a, b, c etc. to the squares in the order that they were presented in the first part of this chapter. This lettering system is maintained in subsequent tables. The total number of squares appearing in Table 2.1 is 36.

The first thing to notice in this dataset is that the bearing angle of the northwest side is restricted to only a few values, and the value that occurs most frequently is overwhelmingly 45 degrees. The northwest side has a bearing angle of 45 degrees for 22 squares, 30 degrees for 5 squares and 60 degrees for 1 square. The west side has a bearing angle of 0 degrees for

Table 2.1: *Squares fitting "square" craters organized by diagonal size (latitude degrees).*

Crater Name	Bearing Angle (°)	Diagonal Size (°)	Longitude (° E)	Latitude (° N)
Elysium Caldera	45	0.3750	146.7734	24.6520
nn10	45	0.4167	279.9978	26.0671
nn2	30	0.4500	141.6979	-8.6918
nn5 (a)	45	0.6667	61.7861	-9.1393
nn5 (b)	0	0.6667	61.7861	-9.1393
E1290	45	0.6667	196.7609	-27.7090
nn12	45	0.7500	47.3657	5.6926
nn14	45	0.8000	46.7522	7.2811
Rauch	45	0.8000	301.8772	21.5504
Albor Caldera	45	0.8333	150.2463	19.0689
nn7	45	0.8333	178.5611	-24.0727
nn9	45	0.9000	240.0336	-24.8813
nn13	45	1.0000	46.8547	6.3758
Verlaine	0	1.0000	64.1236	-9.2386
E1280	45	1.1250	190.7916	-16.3826
Elorza	45	1.2000	304.7904	-8.7500
Mazamba	0	1.2500	290.3470	-27.5226
nn4	45	1.3333	50.8011	10.8689
E999	45	1.6000	65.4849	-7.0000
Teisserenc de Bort (b)	30	1.8750	45.0172	0.4642
nn1	30	2.0000	39.7349	5.2842
Ritchey	45	2.0000	308.9921	-28.4529
nn3	45	2.2500	48.9285	5.5918
McLaughlin	45	2.2500	337.6263	21.8434
Nicholson (b)	45	2.2500	195.5474	0.1504
Nicholson (f)	0	2.2500	195.5958	0.2045
Henry (d)	0	2.4000	23.4774	10.7500
Nicholson (a)	45	2.5000	195.5474	0.1504
Nicholson (c)	30	2.5000	195.5474	0.1504
Nicholson (d)	60	2.5000	195.5474	0.1504
Nicholson (e)	0	2.5000	195.5958	0.2045
Teisserenc de Bort (a)	30	2.6667	45.0172	0.4642
Henry (c)	0	3.3333	23.4774	10.7500
Janssen	45	3.7500	37.5495	2.7183
Henry (a)	45	4.0000	23.4774	10.7500
Henry (b)	0	4.0000	23.4774	10.7500

8 squares. The next thing to notice is that there is more than a 10 fold difference in diagonal size from the smallest to largest crater. The smallest crater (actually a caldera in this instance) has a diagonal size of 0.3750 latitude degrees while the largest crater has a diagonal size of 4.0000 latitude degrees. Now notice that sizes are often repeated. The sizes of 0.6667, 0.8000, 0.8333, 1.0000, 2.0000, 2.2500, 2.5000 and 4.0000 latitude degrees each occur 2 or more times. The most important finding, however, is that diagonal size can be organized into subgroups of values which differ by factors of 2. Thus we have 0.3750 and 0.7500 degrees, 0.4167 and 0.8333 degrees, 0.4500 and 0.9000 degrees and so on. There are a total of 12 subgroups which have members differing by factors of 2 which suggests that there is some sort of underlying pattern to the crater sizes and the squares that fit them.

Music of the Gods

What could be the purpose of these subgroups? The mystery is solved when we look at the diagonal values occurring between 1.0000 and 2.0000 latitude degrees in size. Here we have 1.0000, 1.1250, 1.2000, 1.2500, 1.3333, 1.6000, 1.8750 and 2.0000 degrees. These numbers can be expressed as a ratio between 2 integers. Thus 1.0000 is 1:1, 1.1250 is 9:8, 1.2000 is 6:5, 1.2500 is 5:4, 1.3333 is 4:3, 1.6000 is 8:5, 1.8750 is 15:8 and 2.0000 is 2:1. Sound familiar? Those who have studied music theory will immediately recognize these ratios as the frequency ratios of pure intervals above a fundamental note that are found in the chromatic scale. Hence 1:1 is the perfect unison (P0) interval, 9:8 is the major second (M2) interval, 6:5 is the minor 3rd (m3) interval, 5:4 the major 3rd (M3), 4:3 the perfect 4th (P4), 8:5 the minor 6th (m6), 15:8 the major 7th (M7), and 2:1 the perfect octave (P8) interval. The diagonal sizes below 1.0000 are simply intervals in lower octaves while the sizes above 2.0000 are intervals in higher octaves. Thus 0.6667 degrees is one-half of 1.3333 degrees. The number 1.3333 corresponds to the ratio 4:3 which is a perfect 4th (P4) interval. The value of 0.6667 is therefore a perfect 4th interval in the first octave below the octave ranging from 1.0000 and 2.0000. The number 3.7500 is twice 1.8750 which is the major 7th (M7) interval ratio of 15:8. Thus 3.7500 is a major 7th interval in the octave above the octave between 1.0000 and 2.0000. To complete the translation of diagonal sizes to music intervals, I have assigned all the diagonal values with interval names and the octave number to which they belong in Table 2.2.

This is the first evidence for a Martian system of music. The music intervals represented by the craters in Table 2.2 include all the intervals of our familiar 12 note chromatic scale except two, namely 16:15 for the

Table 2.2: *Music intervals represented by squares fitting "square" craters.*

Crater Name	Diagonal Size (°)	Interval Name	Interval Ratio	Octave Number
Elysium Caldera	0.3750	P5	3:2	-2
nn10	0.4167	M6	5:3	-2
nn2	0.4500	m7	9:5	-2
nn5 (a)	0.6667	P4	4:3	-1
nn5 (b)	0.6667	P4	4:3	-1
E1290	0.6667	P4	4:3	-1
nn12	0.7500	P5	3:2	-1
nn14	0.8000	m6	8:5	-1
Rauch	0.8000	m6	8:5	-1
Albor Caldera	0.8333	M6	5:3	-1
nn7	0.8333	M6	5:3	-1
nn9	0.9000	m7	9:5	-1
nn13	1.0000	P0	1:1	0
Verlaine	1.0000	P0	1:1	0
E1280	1.1250	M2	9:8	0
Elorza	1.2000	m3	6:5	0
Mazamba	1.2500	M3	5:4	0
nn4	1.3333	P4	4:3	0
E999	1.6000	m6	8:5	0
Teisserenc de Bort (b)	1.8750	M7	15:8	0
nn1	2.0000	P0	1:1	+1
Ritchey	2.0000	P0	1:1	+1
nn3	2.2500	M2	9:8	+1
McLaughlin	2.2500	M2	9:8	+1
Nicholson (b)	2.2500	M2	9:8	+1
Nicholson (f)	2.2500	M2	9:8	+1
Henry (d)	2.4000	m3	6:5	+1
Nicholson (a)	2.5000	M3	5:4	+1
Nicholson (c)	2.5000	M3	5:4	+1
Nicholson (d)	2.5000	M3	5:4	+1
Nicholson (e)	2.5000	M3	5:4	+1
Teisserenc de Bort (a)	2.6667	P4	4:3	+1
Henry (c)	3.3333	M6	5:3	+1
Janssen	3.7500	M7	15:8	+1
Henry (a)	4.0000	P0	1:1	+2
Henry (b)	4.0000	P0	1:1	+2

minor 2nd (m2) interval and either 45:32 or 64:45 which are variants of the tritone interval (the interval which is midway in the chromatic scale). One other thing to note is that I have been using the diagonal length of a square as an analogue of sound frequency (cycles per second) which is the unit by which pitch is measured in music. This would put the lower pitched sounds in the smaller craters and the higher pitched sounds in the larger craters. One would expect diagonal size to be more appropriately an analogue of wavelength, so that the smaller craters would represent higher pitched sounds than the larger craters. If such is the case, then the interval names which I have assigned would have to be changed to their complements, and the octave numbers would have to go in descending order rather than ascending order. The complement of an interval is the interval by which it must be raised to bring the sound to 1 octave above the fundamental. For example, a P5 interval must be raised by a P4 interval (3:2 x 4:3 = 2:1) and an M3 interval must be raised by a m6 interval (5:4 x 8:5 = 2:1).

Coordinates of Square Craters

As we have seen, a square offers 4 vertices and a centre for marking coordinate locations. The obvious question then to ask is whether or not "square" craters follow a pattern for coordinate locations such as the systematic marking of every n number of degrees. Alternatively they might be used to locate coordinates having significant meaning such as spiritually important numbers, or integer numbers of degrees. Looking at Table 2.1 again, we can see that in terms of longitude, a high concentration of "square" craters occurs between 37.5° E to 65.5° E. Between these 2 longitudes, square craters occur on average about once every 2.5 degrees. For the rest of the planet, they occur only about once every 22 degrees or about 8.8 times less frequently. In the case of latitudes, 13 or one-half of the craters occur between 10° S and 10° N and only 4 craters occur in the combined bands between 20° S to 10° S and 10° N to 20° N. Such uneven placement on the planet is contrary to the hypothesis that these craters form some sort of grid to mark out a coordinate system. Also, the fact that they are concentrated on the other side of the planet from the major mountains suggests that they were not intended to create any sacred geometry patterns with the mountains. Hence, they must have had a purpose independent of the great mountains.

A clue as to the purpose of the "square" craters may be found in the values of the coordinates of the vertices and centre of the various squares which fit these craters. We have already seen very meaningful coordinate values occurring in the 5 "square" craters and 2 calderas examined so far.

I therefore felt that it was prudent to determine the coordinate values of squares fitting the other 19 "square" craters which I found on the planet between the latitudes of 30° S and 30° N. What I discovered was that every square examined had at least 1 meaningful longitude value and 1 meaningful latitude value. A single longitude value and single latitude value for each square are listed in Table 2.3 to conserve space. The values of the theoretical coordinates are within ±20 seconds of a degree from the actual measurements. Only the Henry (d) square had a deviation of about 1 minute of a degree from the theoretical sacred longitude formula of its northeast and southeast vertices while maintaining the same centre as the other square sizes.

An examination of Table 2.3 reveals that 24 of the 36 latitude values are in terms of sacred formulae involving one of the 6 primary irrational numbers (φ, π, e, $\sqrt{5}$, $\sqrt{3}$, $\sqrt{2}$). Only 12 latitudes are in terms of an integer or integer plus 1/2, but remarkably 5 of these are the integer "7". The longitudes are more in terms of an integer or integer plus 1/2 (19 values) but almost as many of them (17 values) are in terms of a sacred formula demonstrating that longitudes as well as latitudes are quite frequently valued in terms of the irrational numbers rather than only integers or integers plus 1/2 to achieve meaningful numbers. Among the most interesting longitude values are the integers of 1, 5, 30, 54, 90, 100 and 180 and the formulae of Φ, π, and 27φ. The number 54 is one-half the number of degrees between the star points of a pentagram, Φ is the inverse of the golden ratio, and 27 is one-quarter of the number of degrees between star points of a pentagram. If we look at both longitude and latitude values, we find that a total of 27 values or 37.5% make some reference to the pentagram with the numbers 5, 9, 18, 27, 54, 72, φ, Φ (= 1/φ) and $\sqrt{5}$. The numbers 9 and 18 are binary fractions of 36, the number of degrees in a pentagram star point. As just mentioned, the number 27 is one-quarter and the number 54 is one-half the number of degrees between the star points of a pentagram. A total of 8 coordinates use the value π which probably refers to the circle since the ratio of the circumference of a circle to its diameter is π. Six coordinates use 30 or $\sqrt{3}$ which could refer to the equilateral triangle since 30 is one-half the value of its 60 degree angles, and the height of an equilateral triangle is equal to $\sqrt{3}$ times 1/2 the side length. Ten coordinates use $\sqrt{2}$, 4 or 90 which could refer to the square since a square has 4 sides and four 90 degree angles, and the diagonal of a square is equal to $\sqrt{2}$ times the length of one of its sides. All of this suggests that a major function of the squares fitting "square" craters is to generate meaningful numbers used in sacred geometry.

One thing to note is that it is the actual number which is important, not the size in regular degrees. Thus Table 2.3 gives the value 54°° E for the

Table 2.3: *Meaningful coordinates for squares fitting square-shaped craters. Elements include the square centre (c) and the 4 vertices: n=north, s=south, e=east, w=west, nw=northwest, ne=northeast, sw=southwest, se=southeast). P.M. = Prime Meridian and P.L. = Prime Latitude. Coordinates are in regular (°), big (°°) or sacred (°°°) degrees.*

Crater	Element	Longitude	P.M.	Element	Latitude	P.L.
Elysium Caldera	e	1^{oo} W	DPPM	s	$9e^{o}$ N	
nn10	e	26.5^{ooo} E	PMPM	n	27.5^{ooo} N	AMPL
nn2	w	5^{oo} W	CEPM	w	7^{ooo} S	
nn5 (a)	e	$95\sqrt{5}^{oo}$ W	STrPM	n	$(\sqrt{2}/2)^{o}$ S	AMPL
nn5 (b)	sw, nw	77^{oo} W	DMPM	sw, se	7.5^{ooo} S	
E1290	w	$32(\varphi^2)^{ooo}$ W	STrPM	w, e, c	$3(e^2)^{ooo}$ S	
nn12	w	$55\varphi^{oo}$ W	CEPM	n	$3\varphi^{ooo}$ N	
nn14	w	$72\sqrt{2}^{o}$ W	DMPM	w, e, c	$4\varphi^{oo}$ N	
Rauch	n, s, c	Φ^{ooo} E	STrPM	s	26^{oo} N	AMPL
Albor Caldera	n, s, c	π^{o} E	CEPM	n	$9\sqrt{3}^{ooo}$ N	
nn7	w	30° E	DPPM	n	$11\sqrt{2}^{o}$ S	AMPL
nn9	n, s, c	$4\sqrt{2}^{ooo}$ W	PMPM	s	$13\sqrt{3}^{oo}$ S	
nn13	n, s, c	90^{oo} W	DPPM	s	$5\sqrt{5}^{ooo}$ N	AMPL
Verlaine	ne, se	73.5^{oo} W	EMPM	c	$(e^2)^{ooo}$ S	
E1280	n, s, c	$27\varphi^{o}$ E	CEPM	c	$9\varphi^{oo}$ S	
Elorza	w	$50\pi^{o}$ E	CEPM	w, e, c	7^{ooo} S	
Mazamba	c	$5\sqrt{3}^{ooo}$ W	SToPM	c	$9e^{oo}$ S	
nn4	e	85^{oo} W	CEPM	n	$5\pi^{ooo}$ N	AMPL
E999	n, s, c	73.5^{oo} W	DMPM	w, e, c	7° S	
Teisserenc de Bort (b)	e	90^{oo} W	EMPM	e	$(\sqrt{2}/2)^{o}$ N	
nn1	c	95.5^{oo} W	EMPM	n	5^{ooo} N	
Ritchey	w	54^{oo} E	PMPM	s	$9\varphi^{ooo}$ S	
nn3	n, s, c	$30(\varphi^2)^{ooo}$ W	CEPM	s	$4\pi^{o}$ N	AMPL
McLaughlin	e	$15\sqrt{5}^{oo}$ E	STrPM	w, e, c	$12\varphi^{oo}$ N	
Nicholson (b)	w	42^{oo} E	EMPM	n	7.5^{ooo} N	AMPL
Nicholson (f)	c	$58\varphi^{oo}$ W	SToPM	nw, ne	1° N	
Henry (d)	ne, se	$70\sqrt{2}^{ooo}$ W	DPPM	sw, se	18° N	AMPL
Nicholson (a)	w	41^{oo} E	DMPM	s	7° N	AMPL
Nicholson (c)	n	$15\pi^{o}$ E	DPPM	n	$(e/2)^{o}$ N	
Nicholson (d)	s	$12\pi^{ooo}$ E	DPPM	e	$(\sqrt{3}/10)^{o}$ S	
Nicholson (e)	ne, se	43^{oo} E	DPPM	sw, se	$(e/5)^{ooo}$ S	
Teisserenc de Bort (a)	n	180^{oo} W	PCPM	e	$(\varphi/2)^{o}$ N	
Henry (c)	ne, se	98^{ooo} W	EMPM	sw, se	$5\pi^{oo}$ N	AMPL
Janssen	w	100^{oo} W	DMPM	w, e, c	e^{o} N	
Henry (a)	w	$40\pi^{o}$ W	CEPM	s	7^{ooo} N	
Henry (b)	ne, se	221^{ooo} W	SToPM	c	$6\pi^{o}$ N	AMPL

longitude of the west vertex of the square fitting the Richey crater. Even though 54 is in terms of big degrees which is equivalent to 60.75 regular degrees, the number 54 here represents 1/2 the size of the angle in regular degrees between the star points of a pentagram or of the internal angles of a pentagon.

Summary and Conclusions

As a group, the "square" craters provide a huge insight as to how craters are used on Mars to convey important information to the observer. The following is a list of conclusions derived from the study of "square" craters:

1. The circle is not the only geometric shape of a crater on Mars. The "square" shape is found in several instances, and as we shall see, a number of other geometric shapes are also used. This is a significant indicator of artificiality for a large number of Martian craters.

2. A "square" crater is never a perfect square. Its vertices are usually rounded and its sides are incomplete. Sometimes a side can even be missing. However, enough information is always present to construct the square.

3. A critical concept is that it is often the location of one or more components of a fitted geometric shape that is primary rather than the location of a crater's centre for marking a notable site. In the case of a square, the location(s) of 1 or more of the vertices may be the most notable site(s). The important location(s) may not even be on the crater's perimeter or within the crater's boundaries.

4. The square's diagonal seems to be the key parameter for the sizing of a square rather than the side length. The unit of measurement for the diagonal appears to be the distance covered by 1 degree of latitude. Latitude degrees are used rather than longitude degrees since they are constant in distance for all latitudes.

5. Squares fitting square-shaped craters are sized according to music intervals identical to our 12 note chromatic scale, and sizes cover several octaves.

6. Except for squares whose north and south sides are lined up parallel

to lines of latitude, the sides of all fitted squares have a bearing angle which is divisible by 3. The most common bearing angle of the northwest side is 45° in the clockwise direction. It was found in more than 60% of the squares. Squares of this orientation can be used as a compass with the vertices pointing to the 4 cardinal points.

7. "Square" craters can sometimes be fit by several sizes and rotations of squares. The centre of these squares usually is constant but occasionally the centre shifts slightly. A non-constant centre may have been used to make the presence of multiple squares more difficult to detect.

8. Smaller auxiliary craters are often used for alignment purposes for fitting geometric shapes to the main crater. Auxiliary craters can occur within, on or outside the perimeter of a "square" crater. The alignment of a geometric shape can be with the perimeter edge, central peak (centre or edge – see Fig. 1.9) or centre of an auxiliary crater.

9. The squares fitting "square" craters confirm the use of the 8 prime meridians and the 3 basic coordinate systems of 360°, 320°° and 288°°° which were worked out in the first chapter. All produce important numbers for the coordinates of the vertices and centres of squares.

Although all of the "square" craters were found to have meaningful coordinates (integer, integer and 1/2, or sacred formula), they may have had additional purposes other than to express sacred geometry. One possibility is that they could be used to measure altitude for an overhead spacecraft which possessed knowledge of their size and location. The spacecraft would not have to bounce electromagnetic waves off of the planetary surface which would reveal its presence. Another interesting possibility is that they might have been designed to emulate a musical score which encodes a tune such as a greeting song or a sacred melody for an arriving spacecraft.

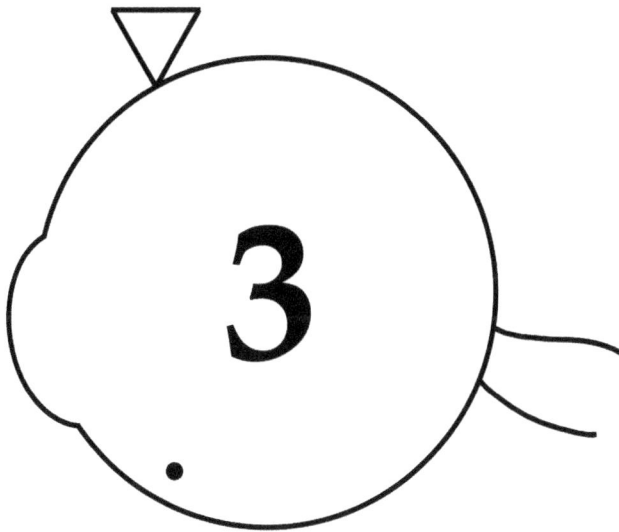

Pentagon Craters

The next shape of crater in the polygon series is the regular pentagon which has 5 equal sides. We have seen how the pentagon shape is found in the Pentagon Pyramid. Now I am going to show you a number of craters that have been deliberately shaped in the form of a regular pentagon. Like the "square" craters, pentagon craters usually have rounded corners and some of the outline can be missing. But there is always enough information to properly construct the full pentagon. There are some craters that have 5 sides but are not regular pentagons. In the regular pentagon, the 5 sides have to be of equal length and each of the internal angles have to equal 108°. This chapter will only consider the craters which can be fit to regular pentagons.

To begin with, let's look at a small unnamed crater (Fig. 3.1) which is 3661 km southwest of Elysium Mons in the Libya Montes region of Mars. It can be well fit by a pentagon which is centred on the coordinates of 85.3013° E 0.6953° N. I gave the crater the name nn37 in my own numbering system. The radius of a circle that passes through the 5 vertices of the pentagon which best fits the crater is 0.1333 latitude degrees, a distance of 7.90 km. I selected this crater because its perimeter follows the pentagon shape almost perfectly and it has amazing coordinates. The first thing to observe is that one of its vertices points exactly due east. Other interesting features of this crater are the coordinates of its centre and

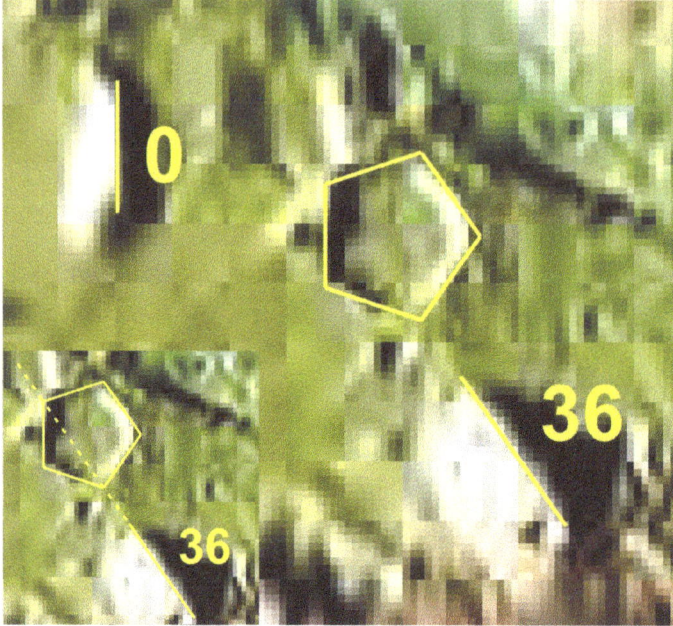

Fig. 3.1: *Pentagon crater nn37 at 85.3013° E 0.6953° N. It is well fit by a pentagon 0.1333 latitude degrees in radius. One of the pentagon's vertices points due east. The linear tops of 2 structures nearby have bearing angles of 0° and 36°. The extended line from the southern structure intersects 2 vertices of the fitted pentagon (inset bottom left). Top of page is north. USGS Astrogeology.*

vertices. Its centre is 54.9965°° W and 49.4969°°° W of the Elysium Mons PM. These numbers are very close to 55 and 49.5. The latitude of its centre and eastern vertex is an impressive 0.6180°° N. I say impressive because 0.6180 is equal to Φ which is the inverse of the golden ratio $\varphi = 1.6180$. This is all the more incredible since the longitude of its centre is exactly $100\Phi = 61.8034°$ W of the Crater Edge PM and $100\varphi = 161.8034°$ W of the Pavonis Mons PM. Thus the longitude difference of 100° between the Crater Edge and Pavonis Mons prime meridians is in golden ratio to the longitude difference of $100\Phi°$ between crater nn37 and the Crater Edge PM, i.e., $100/100\Phi = 1/\Phi = \varphi$. Also the longitude difference of $100\varphi°$ between crater nn37 and the Pavonis Mons PM is in golden ratio to the longitude difference of 100° between the Crater Edge and Pavonis Mons prime meridians, i.e., $100\varphi/100 = \varphi$. The above findings provide extremely strong evidence for the existence of the CEPM and the PMPM.

The latitudes of the other vertices of the pentagon which fit crater nn37 also have meaningful numbers. The latitude of the northern vertex (0.6577°°° N) is close to $\varphi^2/4 = 0.6545$. Besides their latitude of $\Phi°°$ N, the eastern vertex and pentagon centre also have a latitude of 0.5562°°° N which is close to $\sqrt{5}/4 = 0.5590$. The southern vertex is 8.6681° N and 6.9345°°° N (both with reference to the Arsia Mons Prime Latitude). These values are close to the symmetrical numbers $5\sqrt{3} = 8.6603$ and $4\sqrt{3} = 6.9282$. The southwest vertex has a latitude of 7.7480°° N (AMPL) which is close to $10\sqrt{3}/\sqrt{5} = 7.7460$. Finally, the northwest vertex has the latitude of

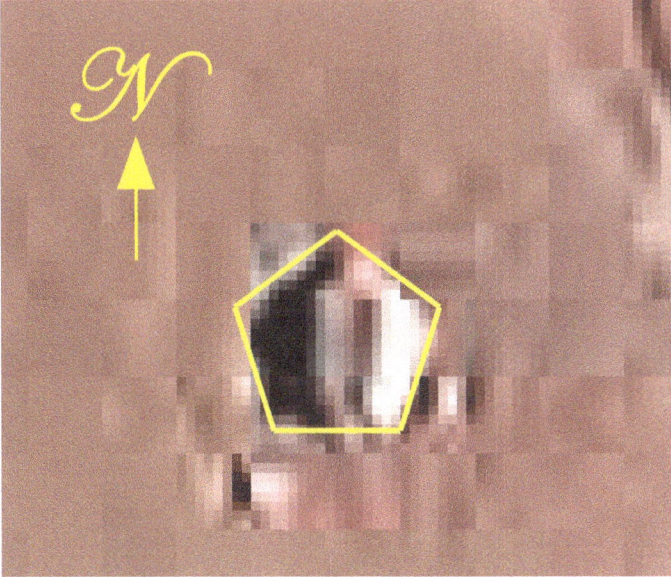

Fig. 3.2: *Pentagon crater nn46 just south of the Coprates Chasma. It has a vertex which points due north. Its radius is 0.125 latitude degrees. USGS Astrogeology.*

0.7737° N and 0.6190°°° N, numbers which are close to √3/√5 = 0.7746, and to Φ = 0.6180 in sacred degrees instead of the big degrees used for the crater centre latitude.

Also of interest in Fig. 3.1 are 2 large structures close to the crater which have linear top edges. The bearing angles of these top edges are 0° and 36° which suggests that the structures are artificial. The angle of 36° is the size of the angle in a pentagram star point. Since the pentagon geometric shape is contained in the pentagram, the bearing angle of 36° would imply that the structure to the southeast is related to the nn37 pentagon crater. This is reinforced by the finding that if its top edge line is extended to the northwest, it intersects perfectly with the south and northwest vertices of the pentagon fitting the nn37 crater (see inset), and forms an isosceles triangle in the pentagon with base angles equal to 36°.

Because the pentagram encodes φ in many ways, the finding of Φ and φ in the coordinate values of the centre of the pentagon which fits crater nn37 suggests that this very remarkable crater was intended to honour the golden mean and the pentagram. It would not surprise me if it had been called the φ Crater or the Golden Mean Crater in ancient times.

The pentagon which fits the crater (nn46 in my numbering system) in Fig. 3.2 is located at 291.7024° E 14.7265° S which puts it on the opposite side of the planet from the previous crater. It has a radius of 0.125 latitude degrees = 7.41 km, and one of its vertices points due north. The pentagon centre has a latitude of 13.0902°° S where the numerical value is equal to $5\varphi^2$. The latitude of the western and eastern vertices is 11.7502°°° S, a

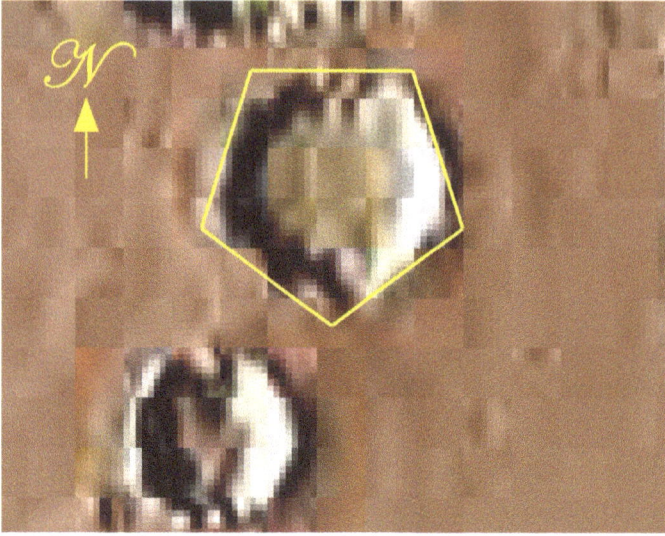

Fig. 3.3: *The Wallula Crater can be fit to a regular pentagon which has a radius of 0.25 latitude degrees. The fitted pentagon has a vertex pointing due south. The centre of the pentagon and its southern vertex have a longitude equal to $\pi^{\circ\circ\circ}$ E (Sharonov Triangle PM). The latitude of its north side is $\pi/2^{\circ\circ\circ}$ S (AMPL). USGS Astro-geology.*

value which would be close to an integer in a system with 4 times the number of degrees making up the 288 degree system. The longitude of the southwestern vertex is $6\sqrt{2} = 8.4853^{\circ\circ}$ W of the Sharonov Triangle PM. The eastern vertex is $44.7206°$ E and $35.7765^{\circ\circ\circ}$ E of the Pavonis Mons PM. These numbers are very close to $20\sqrt{5} = 44.7214$ and $16\sqrt{5} = 35.7771$.

The Wallula Crater can be fit to a pentagon which has a vertex that points due south (Fig. 3.3). The coordinates of the pentagon centre are $305.0317°$ E $10.2653°$ S and the radius size is 0.25 latitude degrees or 14.82 km. The perimeter of this crater does not follow the pentagon shape perfectly but linear sections of the perimeter touch the pentagon outline. Its centre has a latitude of $1.7326^{\circ\circ\circ}$ S with reference to the Arsia Mons Prime Latitude which is very close in numerical value to $\sqrt{3} = 1.7321$. The north side of the pentagon sits at $\pi/2 = 1.5708^{\circ\circ\circ}$ S (AMPL). This latitude can also be expressed as $8.9450^{\circ\circ}$ S with reference to the equator which has a numerical value very close to $4\sqrt{5} = 8.9443$. With regards to longitude, the centre of the fitted pentagon and its southern vertex is $\pi = 3.1416^{\circ\circ\circ}$ E of the Sharonov Triangle PM. This longitude and the latitude of $\pi/2^{\circ\circ\circ}$ S (AMPL) for the north side of the pentagon suggest that this pentagon was constructed mainly to honour the value of π.

The Pentagon Compass

I am now going to show you a crater (Fig. 3.4) that serves as a compass which points to the 4 cardinal directions of north, east, south and west. As the crater is unnamed, I originally assigned the designation of nn57 in my

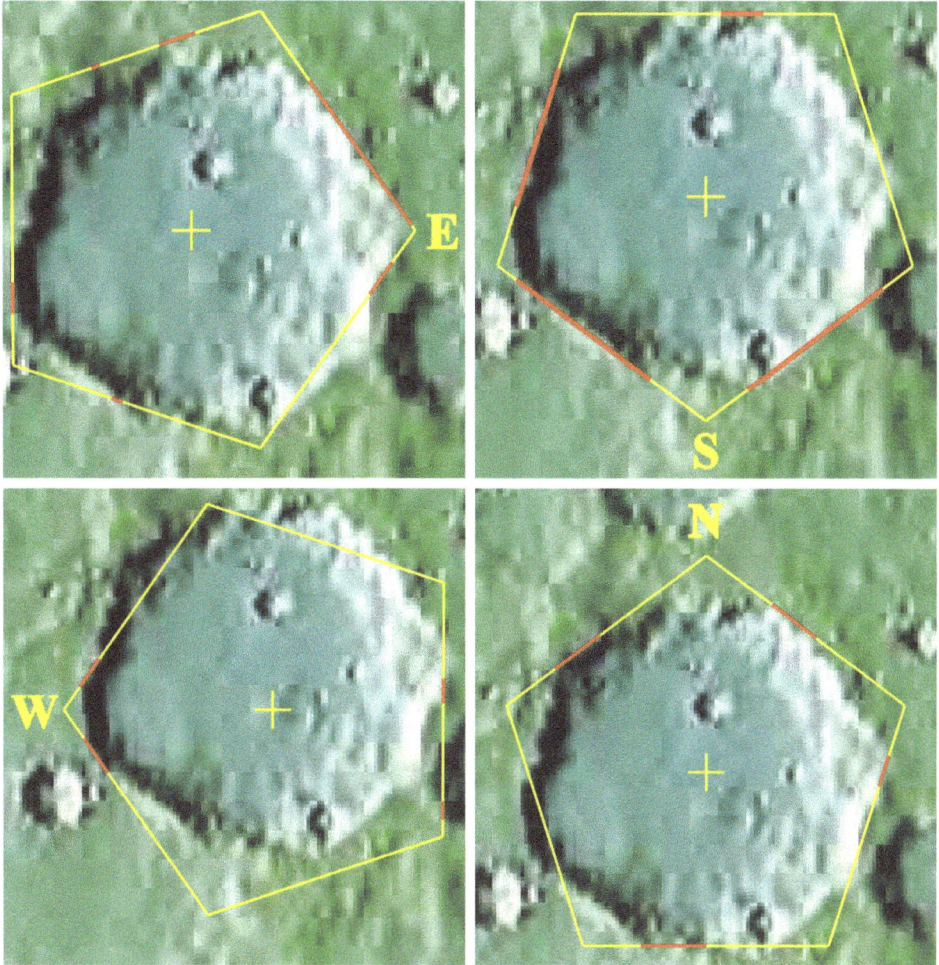

Fig. 3.4: *The Compass Crater can be fit to 4 distinct pentagons. Each pentagon points to a different cardinal point of the compass and has its own unique centre. Many of the pentagon vertices and centres have meaningful coordinates. USGS Astrogeology.*

own numbering system but I also called it the Compass Crater for reasons which will become apparent below. The centre of this crater is difficult to locate precisely since its perimeter is irregularly shaped but its approximate coordinates are 352.2° E 26.2° N. Its diameter is about 64 km so it is a fairly large crater which makes it easy to spot from a high altitude. The upper left picture in Fig. 3.4 shows that a perfect pentagon which has a radius of 0.6 latitude degrees or 35.56 km can be fit to several features of the crater perimeter. All sides are aligned to linear sections of the crater perimeter or to the edges of outward projections of the crater perimeter. This pentagon has a vertex which points due east so it can be used to precisely indicate the eastern direction. The pentagon centre

(352.1724° E 26.2500° N) and its eastern vertex lie at a latitude of 21.0000°°°
N. Note that 21 is a Fibonacci number. The longitude of the centre is
205.0000° E or 164.0000°°° E from the Elysium Mons PM, 204.0000° E from
the Dagger Midline PM, 105.0000° E or 84.0000°°° E from the Pavonis
Caldera PM and 51.0000° E from the Sharonov Tower PM.

If we rotate the pentagon by 18 degrees in the clockwise direction and
relocate the centre slightly to 352.2196° E 26.2250° N we obtain the
pentagon shown in the upper right picture of Fig. 3.4. The pentagon now
aligns with different linear segments of the crater perimeter than with the
previous pentagon. Only the northeast side does not align to any part of
the crater perimeter. One of the vertices now points due south. The
latitude of the southern vertex is 20.5000°°° N. The longitude of the
eastern vertex is 204.7500° E or 182.0000°° E of the Dagger Peak PM,
105.7500° E or 94.0000°° E of the Pavonis Mons PM, and 51.7500° E or
46.0000°° E of the Sharonov Triangle PM.

Rotating the pentagon once again by 18 degrees in the clockwise
direction and relocating the centre slightly to 352.2362° E 26.1250° N
creates a pentagon which fits small linear sections of the crater perimeter
which are at different locations than for the previous 2 positions of the
pentagon (Fig. 3.4, lower left). Three sides are aligned whereas the
northeast and southeast sides are unaligned. The east side, besides
aligning with a short linear segment of the crater perimeter, also aligns
with a step in the perimeter of an adjacent crater. This is reminiscent of
the role played by auxiliary craters which were found to be associated
with the "square" craters of the previous chapter. This latest pentagon has
a vertex which points due west. The midpoint between the western vertex
and the middle of the eastern side of the pentagon has an interesting
longitude. It is the same as the centre of the pentagon above which
pointed due east, namely 205.0000° E or 164.0000°°° E of the Elysium
Mons PM, 204.0000° E of the Dagger Midline PM, 105.0000° E or
84.0000°°° E of the Pavonis Caldera PM and 51.0000° E of the Sharonov
Tower PM.

Finally, rotating the pentagon by another 18 degrees in the clockwise
direction and placing its centre at 352.2175° E 26.2354° N produces a
pentagon which fits different linear sections of the crater perimeter than
the 3 previous versions and now has a vertex which points due north (Fig.
3.4, lower right). All sides are aligned with the crater perimeter except the
southwest side. The latitude of the southwest and southeast vertices of the
pentagon is 25.75000° N. More interesting is the latitude of the northern
vertex at 26.8354° N which is numerically very close to $12\sqrt{5} = 26.8328$. The
longitude of the east vertex is the same as the longitude of the east vertex
for the south pointing pentagon above even though the centres have

different longitudes. This is because the latitude of the east vertex of the north pointing pentagon is greater than that for the east vertex of the south pointing pentagon. This causes the longitude of the east vertex of the north pointing pentagon to shift to the exact value of the east vertex of the south pointing pentagon and shows great genius in the placement of the 2 pentagons. The longitude of the east vertex is 204.7500° E or 182.0000°° E (Dagger Peak PM), 105.7500° E or 94.0000°° E (Pavonis Mons PM), and 51.7500° E or 46.0000°° E (Sharonov Triangle PM).

All of the 4 rotations of the pentagon fit different linear segments of the crater perimeter. In 3 of the rotations, 1 or more sides out of 5 do not align with linear segments of the perimeter but enough alignment information is present to construct the complete pentagon. All fitted pentagons have a vertex which points to one of the 4 cardinal points of the compass. Hence, this crater could have been used as a compass by craft flying overhead. The east, south and north pointing pentagons also mark out the locations of important coordinates either with their centres or with one of their vertices. The west pointing pentagon marks an important longitude with the midpoint between its west vertex and its eastern side. Although each of the pentagons has a slightly different location for its centre, this does not prevent them from being used as a compass. The presence of whole integers, half integers, quarter or three quarter integers in the values of the coordinates of their vertices suggest the use of double and quadruple degree systems which would permit the Martians to divide the globe much more finely than we do with our 360 degree system.

The Curie Crater

The first thing that strikes you about the Curie Crater is its massive size (Fig. 3.5). I measured its diameter to be about 120 km. Although not as large as the Henry Crater, the Curie Crater spans 2 degrees of longitude and 2 degrees of latitude and covers more than 10,000 square km of the planet's surface. Not bad for a single crater. I initially assumed that the crater was named after Marie Curie, the 2 time winner of the Noble prize. I was somewhat surprised to discover that it was actually named after her husband, Pierre Curie, a very meritorious scientist himself. Perhaps those responsible for assigning crater names are simply waiting until they find an even more impressive site to name after Marie. After all she won her Nobel prizes in both chemistry and physics. A quick scan of all the Martian crater names failed to turn up any females who were honoured in this way, so Marie and other famous female scientists and female science fiction writers may have to wait a very long time.

When I examined the Curie Crater more closely, I noticed that rather

Fig. 3.5: *The Curie Crater has large linear segments in its perimeter especially for its northwest and southwest sides. It is named after Pierre Curie, the husband of the more famous Marie Curie. USGS Astrogeology.*

than being round like traditional craters, the perimeter of this crater was largely composed of straight line segments, some of them long enough to cover more than 5% of the entire perimeter length. It sort of looked a bit like a square, but the opposite sides were not parallel so it could not be a square. Then it struck me. This crater actually has the shape of a regular pentagon! The reason you don't notice it at first is because much of the perimeter is difficult to locate precisely especially on the eastern side. After several attempts at fitting a regular pentagon to the crater, I finally settled on the fit shown in Fig. 3.6. The reasons for my selection become apparent when you look at the coordinates of the centre, the actual size of the pentagon and, of course, the excellent fit to major linear regions of the crater perimeter. The fit to the northeastern side is to changes in colouration in the crater wall rather than to the crater perimeter.

The coordinates of the pentagon centre are 355.1724° E 28.7500° N. At first glance, this does not seem too impressive. The latitude is an integer plus 3/4 of an integer, suggesting a 1440 degree system. When you convert it to sacred degrees, however, it becomes a pure integer, namely, 23.0000°°° N. Its longitude is even more interesting when it is referenced to prime meridians set up by the Martian architects. Thus the centre is located at 208.0000° E of the Elysium Mons PM, 207.0000° E or 184.0000°° E of the Dagger Midline PM, 108.0000° E or 96.0000°° E of the Pavonis Caldera PM, and 54.0000° E or 48.0000°° E of the Sharonov Tower PM.

Fig. 3.6: *An excellent fit to the Curie Crater can be achieved with a pentagon having a radius of one latitude degree. All sides align to linear segments of the crater perimeter except the northeast side which aligns to colour changes in the crater wall. The lower yellow cross marks the centre of the pentagon and the upper yellow cross marks the midpoint between the northern vertex and south side of the pentagon. USGS Astrogeology.*

The numbers 54 and 108 really caught my eye. The number 108 is exactly the number of degrees contained in each of the internal angles of a regular pentagon. The number 54 is half of this number of degrees. So here we have the longitudes from 2 prime meridians which are unmistakable symbols of a regular pentagon. How appropriate! This finding establishes beyond a reasonable doubt the artificiality of the location of the Curie Crater, and by extension, the artificiality of the crater itself. This, of course, is on top of the fact that the crater has the shape of a pentagon rather than a circle.

So in terms of longitude, the Sharonov Tower sits exactly halfway between the Pavonis Mons Caldera and the centre of the Curie Crater pentagon. This suggests that the Curie Crater might be marking another pair of prime meridians but more investigation is required to establish whether or not this is so. The south side of the pentagon is at a latitude of 22.3528° N which is numerically close to $10\sqrt{5} = 22.3607$. The most interesting latitude, however, seems to be that of the midpoint between the south side and northern vertex of the pentagon which is marked by the upper yellow cross in Fig. 3.6. It is 36.9451° N (AMPL) and 29.5560°°° N (AMPL) which are numerically very close to $5e^2 = 36.9453$ and $4e^2 = 29.5562$. Note that 5 is the number of equal sides in a regular pentagon.

The size of the pentagon is also remarkable. Its radius is exactly equal

Fig. 3.7: *If the radius of the pentagon is increased to 1.25 latitude degrees, the pentagon aligns with linear portions of the crater perimeter on the east sides and the outside edge of the crater apron on the southwest side. The south side passes through the central peak of an auxiliary crater southwest of the Curie Crater. The upper cross is midway between the south side and the northern vertex. USGS Astrogeology.*

to one degree of latitude. This puts the northern vertex at a latitude of 29.7500° N which does not translate into an integer number of sacred degrees. However, it has exactly the same longitude coordinates as the centre of the pentagon. The southwest and northwest sides of the pentagon align perfectly with lengthy linear sections of the crater perimeter. The same can be said for the southeast side which aligns with a shorter length of crater perimeter not far from the southeast vertex. The south side aligns with a very short linear section of the perimeter. The northeast side seems to align with the edges of intermediate regions of the crater wall which show up as colour changes. All of this adds up to a very credible fit of the pentagon to a sizeable part of the Curie Crater perimeter.

It would seem that the Curie Crater is not restricted to fitting a single pentagon. If we increase the size of the original pentagon by a factor of 1.25, we get the pentagon shown in Fig. 3.7. The southwest side of this pentagon hugs a lengthy section of the outside edge of the crater apron. Its south side passes through the central peak of a small crater to the southwest of the Curie Crater. This smaller crater is reminiscent of the auxiliary craters that we have seen with the square craters in the previous chapter. The northeast side runs along a linear section of the crater perimeter close to the northern vertex. Except for a small overflow region, the southeast side aligns with the furthest outreaches of the crater perimeter along the lower half of the side. The most interesting feature of this pentagon is that the latitude of the northern vertex is exactly at 30.0000° N or 24.0000°°° N. The number 30 appears to be a very important value for the Martian architects and we have seen it in the bearing angles of important structures such as in the bearing angle of the northwest side of 5 of the squares in the previous chapter. Its importance may derive from the fact that it is 1/2 the size of the angles of an equilateral triangle. The northern vertex is also 38.0996° N (AMPL) which is numerically close to 22√3 = 38.1051. The eastern and western vertices have a latitude of 29.1362° N and 25.8989°° N. These values are close to 18φ = 29.1246 and 16φ = 25.8885.

Characteristics of Pentagon Craters

In order to get a handle on what purpose the pentagon-shaped craters served, I analyzed 41 craters whose perimeters provided enough linear segment information on at least 3 sides to fit a regular pentagon. For this database, I tried to find all the pentagon craters between the latitudes of 30° S and 30° N, the latitude boundaries of my MOLA Mercator maps. Table 3.1 lists these craters in order of size. All the sizes are given in terms of latitude degrees for the pentagon radius. This is the metric that I found to give the most meaningful description of pentagon size. You will notice that many of the sizes occur more than once. Two of the craters were fit by multiple pentagons: 4 pentagons of the same size but different rotations for crater nn57 (the Compass Crater), and 2 pentagons of different size for the Curie Crater, bringing the total number of pentagons in Table 3.1 up to 45.

Remarkably, all of the "pentagon" craters were found to be well fit to a pentagon where one of the vertices pointed to one of the 4 cardinal points of the compass, namely east, south, west or north. The accuracy with which the degree of rotation of the pentagon could be determined was within plus or minus 1 degree. As there are 18 degrees between compass-pointing rotations for pentagons, the odds of a single pentagon having a vertex pointing to a cardinal point is therefore 1 in 9 (i.e., there is a 2 degree resolution for the amount of rotation). The odds of having all 41 "pentagon" craters fit with a compass-pointing pentagon is much less than 1 in one trillion trillion. Hence, the only reasonable conclusion is that these craters have been artificially constructed. The compass direction most often pointed to for the pentagons in Table 3.1 is east (n = 23 or 51%) followed by north (n = 13 or 29%), then south (n = 6 or 13%) and west (n = 3 or 7%). If this distribution between the cardinal directions occurred under random conditions, the probability of having 23 craters pointing

Table 3.1: *Coordinates of the centres of pentagons fitting pentagon craters. Each pentagon has a vertex pointing directly to north (N), east (E), south (S) or west (W). The pentagons are listed according to size (radius in terms of latitude degrees).*

Crater Name	Radius To	Longitude (°)	Latitude (°E)	(°N)	Crater Name	Radius To	Longitude (°)	Latitude (°E)	(°N)
nn47	E 0.1172	300.3242	-22.5287		nn23	E 0.3333	28.6608	-11.4975
E635	E 0.1250	62.1173	17.9265		nn58	E 0.3750	354.6990	4.1799
nn46	N 0.1250	291.7024	-14.7265		nn36	N 0.3750	60.8256	12.7062
nn37	E 0.1333	85.3013	0.6953		nn40	N 0.4000	169.1728	-28.3979
ET638	E 0.1406	303.4084	-8.6163		E966	S 0.4167	58.2126	-1.6390
nn32	E 0.1563	42.7744	-27.4529		nn35	S 0.4167	59.9004	-2.5156
nn21	E 0.1563	28.9638	20.5332		nn53	E 0.4688	344.7776	-1.1722
nn44	E 0.1563	188.8921	-27.4308		nn22	E 0.5333	28.6724	8.5000
nn43	N 0.1667	178.3172	-21.2413		E836	E 0.5625	30.6297	5.9715
nn50	E 0.1667	310.9290	-18.4816		nn51	W 0.5625	314.4899	-23.7137
nn30	E 0.2000	40.6724	28.5180		nn57	E 0.6000	352.1724	26.2500
Timaru	N 0.2000	337.6724	-25.2566		nn57	S 0.6000	352.2196	26.2250
nn41	N 0.2083	175.1928	-23.6449		nn57	W 0.6000	352.2362	26.1250
E977	S 0.2250	57.8257	0.8786		nn57	N 0.6000	352.2175	26.2354
nn54	N 0.2250	349.7646	-9.9429		nn28	E 0.6000	34.8332	27.4294
nn33	E 0.2344	49.8092	-11.3368		nn52	E 0.7500	326.2203	9.8444
nn34	W 0.2344	51.5773	27.9509		nn55	N 0.8000	349.3110	24.3558
E598	N 0.2344	30.5908	17.2481		E788	E 0.8333	27.8801	28.1250
Wallula	S 0.2500	305.0317	-10.2653		nn26	N 0.9000	26.0267	-23.0240
nn29	E 0.2500	39.8979	-12.3531		Curie	N 1.0000	355.1724	28.7500
nn31	S 0.2813	44.1441	-21.6340		Pasteur	E 1.0667	24.6724	19.2355
nn27	E 0.3000	33.6047	-23.0996		Curie	N 1.2500	355.1724	28.7500
nn48	E 0.3125	308.9998	-23.0784					

east is only 0.000148 or about 1 in 6750. Similarly, the probability of only 3 of them pointing to the west is 0.000300 or about 1 in 3300. This is highly suggestive that the distribution amongst the cardinal directions was also the result of intelligent design.

When the distribution of the 41 "pentagon" craters is examined across longitude, it is found that 33 of them (80.5%) are located between 60° W to 60° E (i.e., 300° E to 60° E, moving east). In contrast there are no "pentagon" craters between 90° E to 165° E and between 190° E to 290° E. Except for the region containing Apollinaris Mons, the regions of longitude which contain the biggest mountains on Mars are those containing no "pentagon" craters. There is also an uneven distribution of

Table 3.2: *Chromatic scale intervals for pentagon fits to "pentagon" craters.*

Crater Name	Radius (°)	Inter-val Ratio	Inter-val Name	Octave No.	Crater Name	Radius (°)	Inter-val Ratio	Inter-val Name	Octave No.
nn47	0.1172	15/8	M7	-4	nn23	0.3333	4/3	P4	-2
E635	0.1250	1/1	P0	-3	nn58	0.3750	3/2	P5	-2
nn46	0.1250	1/1	P0	-3	nn36	0.3750	3/2	P5	-2
nn37	0.1333	16/15	m2	-3	nn40	0.4000	8/5	m6	-2
ET638	0.1406	9/8	M2	-3	E966	0.4167	5/3	M6	-2
nn32	0.1563	5/4	M3	-3	nn35	0.4167	5/3	M6	-2
nn21	0.1563	5/4	M3	-3	nn53	0.4688	15/8	M7	-2
nn44	0.1563	5/4	M3	-3	nn22	0.5333	16/15	m2	-1
nn43	0.1667	4/3	P4	-3	E836	0.5625	9/8	M2	-1
nn50	0.1667	4/3	P4	-3	nn51	0.5625	9/8	M2	-1
nn30	0.2000	8/5	m6	-3	nn57	0.6000	6/5	m3	-1
Timaru	0.2000	8/5	m6	-3	nn57	0.6000	6/5	m3	-1
nn41	0.2083	5/3	M6	-3	nn57	0.6000	6/5	m3	-1
E977	0.2250	9/5	m7	-3	nn57	0.6000	6/5	m3	-1
nn54	0.2250	9/5	m7	-3	nn28	0.6000	6/5	m3	-1
nn33	0.2344	15/8	M7	-3	nn52	0.7500	3/2	P5	-1
nn34	0.2344	15/8	M7	-3	nn55	0.8000	8/5	m6	-1
E598	0.2344	15/8	M7	-3	E788	0.8333	5/3	M6	-1
Wallula	0.2500	1/1	P0	-2	nn26	0.9000	9/5	m7	-1
nn29	0.2500	1/1	P0	-2	Curie	1.0000	1/1	P0	0
nn31	0.2813	9/8	M2	-2	Pasteur	1.0667	16/15	m2	0
nn27	0.3000	6/5	m3	-2	Curie	1.2500	5/4	M3	0
nn48	0.3125	5/4	M3	-2					

"pentagon" craters across latitude, with 11 occurring between 10° S and 10° N, 10 between 10° to 20° latitude north and south, and 20 (49%) occurring between 20° to 30° north and south.

What is most interesting is that you can express all of the pentagon sizes in terms of pure integer ratios which correspond to the 12 intervals of the chromatic music scale (Table 3.2), just as was found for the diagonal size of "square" craters in the previous chapter. In order to do this you simply have to divide the radius size by the size of the fundamental note in the octave to which it belongs. I assigned the value of 1 latitude degree to the fundamental note of octave # 0. The fundamental note of the first octave below this (i.e., octave # -1) is obtained by dividing 1 latitude degree by 2 = 0.5. The value of the fundamental for the next octave below this is obtained by dividing 0.5 by 2 = 0.25, and so on for the consecutive lower octaves Thus for the 3rd and 4th octaves below octave # 0 the values of the fundamentals are 0.125 and 0.0625 respectively. As an example for interval calculation, the first entry in Table 3.2 has a radius size of 0.1172 latitude degrees. This value lies between the values for the fundamentals for the 3rd and 4th octaves below octave # 0 so it will be divided by the lower fundamental value. Hence, the interval size is 0.1172/0.0625 = 1.875 which is equal to the integer ratio of 15/8 for the major 7th interval.

Like the squares, the pentagon sizes cover 5 octaves. Eleven of the notes of the chromatic scale are present 3 to 6 times. Only the tritone interval is absent. If we were to measure intervals assuming crater size to be an analogue of wavelength (smaller craters having higher frequencies) rather than of frequency (smaller craters having lower frequencies), then the "notes" would be the complements (inverses or inversions) of the ones listed in Table 3.2 and the octaves would go in descending rather than ascending order. Thus a perfect 5th would then be a perfect 4th, a major 3rd would be a minor 6th and so on.

It was found that all of the fitted pentagons had at least one element (vertex, centre or midpoint between the compass-pointing vertex and the opposite side) which was located at a meaningful longitude coordinate and another or the same element which was located at a meaningful latitude coordinate (Table 3.3). The difference between the theoretical and actual coordinates does not exceed ±0.005 degrees for any of the values listed in Table 3.3. There are many other instances that I could not put into Table 3.3 due to space limitations. The values that I considered meaningful were those that were an integer, integer and one-half, or an integer multiple or integer division of π, π^2, φ, φ^2, e, e^2, $\sqrt{5}$, $\sqrt{3}$ or $\sqrt{2}$. Only the latitude, $1/(\varphi\sqrt{3})^{\circ\circ\circ}$ N, of the south vertex of the west-pointing pentagon fitting the nn57 crater involved 2 irrational numbers. Many of the coordinate values seem to be a reference to the pentagram. The number 9 appears in 4 of the latitudes and the number 18 in 3 latitudes and 1 longitude, the number 54 in 5 longitudes, the number 108 in 1

Table 3.3: *Meaningful coordinates for pentagons fitting pentagon-shaped craters. Pentagon elements include the pentagon centre (c), the 5 vertices (the names depend on the pentagon's orientation: n = north, s = south, e = east, w = west, nw = northwest, ne = northeast, sw = southwest, se = southeast) and the midpoint between the compass-pointing vertex and the opposite side (m). P.M. = Prime Meridian and P.L. = Prime Latitude. Note that coordinates are in terms of regular degrees (°), big degrees (°°) or sacred degrees (°°°).*

Crater	Element	Longitude	P.M.	Element	Latitude	P.L.
nn47	n, s	$(\varphi/2)°$ W	SToPM	s	$9\sqrt{5}°$ S	
E635	n, s	$25e^{ooo}$ W	CEPM	nw	$18°$ N	
nn46	sw	$6\sqrt{2}^{oo}$ W	SToPM	c	$5(\varphi^2)^{oo}$ S	
nn37	c	$100\Phi°$ W	CEPM	c, e, m	Φ^{oo} N	
ET638	c	$\sqrt{5}°$ E	SToPM	s	7^{ooo} S	
nn32	e	$54\sqrt{3}^{oo}$ W	DMPM	n	$15\varphi^{oo}$ S	
nn21	m	105^{oo} W	CEPM	s	$6e^{ooo}$ N	
nn44	s	$12e^{ooo}$ E	DMPM	sw	$9e^{oo}$ S	
nn43	se	25^{ooo} E	EMPM	w	$5(\varphi^2)°$ S	AMPL
nn50	e	$10°$ E	STrPM	n	$7(\varphi^2)°$ S	
nn30	c	86^{ooo} W	DMPM	nw	$18\sqrt{2}^{oo}$ N	
Timaru	c, n, m	$36.5°$ E	SToPM	se, sw	$10\sqrt{3}°$ S	AMPL
nn41	e	$15\varphi^{oo}$ E	DPPM	m	21^{oo} S	
E977	w	$36\sqrt{5}^{oo}$ E	DMPM	e, w	$\varphi/2°$ N	
nn54	w	43^{oo} E	SToPM	se, sw	9^{oo} S	
nn33	c	$54\varphi^{oo}$ W	DPPM	sw	3^{oo} S	AMPL
nn34	sw, nw	$30(e^2)^{oo}$ W	SToPM	c, e, m	$10\sqrt{5}^{ooo}$ N	
E598	se	103.5^{oo} W	EMPM	e, w	$10\sqrt{3}°$ N	
Wallula	c, s, m	π^{ooo} E	STrPM	ne, nw	$\pi/2^{ooo}$ S	AMPL
nn29	m	209^{ooo} E	SToPM	sw	10^{ooo} S	
nn31	e	$33\pi°$ E	DPPM	ne, nw	$7e^{oo}$ S	
nn27	c	$113.5°$ E	CEPM	c, e, m	$15°$ S	AMPL
nn48	n,s	$8°$ W	STrPM	s	$12\sqrt{3}°$ S	
nn23	c	$53\sqrt{5}°$ E	EMPM	c, e, m	e^{ooo} S	AMPL
nn58	m	$184.5°$ E	EMPM	nw	10^{ooo} N	AMPL
nn36	e	$54\sqrt{2}^{oo}$ W	CEPM	m	$7\varphi^{oo}$ N	
nn40	e	18^{ooo} E	CEPM	w	$9\pi°$ S	
E966	e	$60\pi°$ E	PMPM	e, w	$\sqrt{2}^{ooo}$ N	
nn35	w	$12(e^2)°$ W	DMPM	c	$\sqrt{5}^{oo}$ S	
nn53	n, s	35^{ooo} E	SToPM	s	$\varphi°$ S	
nn22	c	218^{ooo} W	SToPM	c, e, m	$8.5°$ N	
E836	sw, nw	$118°$ E	DMPM	s	$2e°$ N	
nn51	se, ne	$48\sqrt{2}°$ E	PMPM	ne	$12\sqrt{3}^{oo}$ S	
nn57	c	84^{ooo} E	PCPM	c, e, m	21^{ooo} N	
nn57	e	94^{oo} E	PMPM	s	20.5^{ooo} N	
nn57	m	84^{ooo} E	PCPM	s	$1/(\varphi\sqrt{3})^{ooo}$ N	
nn57	e	94^{oo} E	PMPM	n	$12\sqrt{5}°$ N	
nn28	n, s	100.5^{oo} W	DPPM	n	$28°$ N	
nn52	n	$13\sqrt{3}^{oo}$ E	STrPM	nw	$13\sqrt{2}°$ N	AMPL
nn55	sw	160.5^{ooo} E	DMPM	n	$10\sqrt{5}^{oo}$ N	
E788	n, s	$120°$ E	DMPM	c, e, m	25^{oo} N	
nn26	se	108^{oo} W	DPPM	se, sw	$7\sqrt{5}°$ S	AMPL
Curie	c, n, m	$54°$ E	SToPM	m	$5(e^2)°$ N	AMPL
Pasteur	c	98^{ooo} W	EMPM	n	18^{oo} N	
Curie	c, n, m	$54°$ E	SToPM	n	$30°$ N	

longitude, and the number 36 in 1 longitude. All of these numbers are the sizes of angles in the pentagram or binary factors of the angles. The number 5 occurs in 3 latitudes, the number 10 in 6 latitudes and 1 longitude, and the number 25 in 1 latitude and 2 longitudes. These numbers are the number, or multiples of the number, of star points in the pentagram. There are also coordinates composed of just a single irrational number, i.e., e, Φ, φ, $\sqrt{2}$, $\sqrt{5}$ or π. The occurrence of such meaningful coordinates strongly suggests that the size and position of the "pentagon" craters were artificially chosen to reflect sacred geometry.

Also noteworthy is the appearance of the number 7 in 5 of the latitude values in Table 3.3, once as a pure integer, and 4 times as a multiplier of an irrational number or the square of an irrational number. This is reminiscent of the 5 occurrences of the number 7 as a pure integer in the latitudes of the elements of the squares fitting "square" craters in Table 2.3.

Summary and Conclusions

Once again, we have solid proof of artificiality in the landscape of Mars. Craters are supposed to be round, not "pentagonal". The regular pentagon is a complex geometric shape that could not be caused by the impact of a meteorite. Even if there were geological anomalies at the site of impact which would favour this shape, the chances of this occurring for even a single crater are extremely low, let alone for 41 different craters. Also the placement of the centre of the pentagons which fit the Curie Crater at 54° E of the Sharonov Tower PM and 108° E of the Pavonis Mons Caldera PM is a sure sign of intelligent design since 108 degrees is the size of each of the internal angles of a pentagon and 54 is one-half of this value. And the odds of having all fitted pentagons pointing to a cardinal point of the compass under random conditions are so low that the artificiality hypothesis is confirmed beyond a reasonable doubt.

"Pentagon" craters form a special subset in the Martian crater landscape. Because they all have a vertex which points either north, south, east or west, they may have been intended to act as compasses for overflying spacecraft. But the fact that more than half of them point due east suggests that they might have served a spiritual purpose. Most of the temples that have been constructed on planet Earth face due east, even Christian churches which were often built on the sites of ancient temples. It is suggestive of a form of sun worship since the sun rises in the east.

The finding that they are sized in accordance to the magnitude of intervals in the chromatic music scale suggests that these craters may have a musical role as well as a spiritual role. Do they simply pay homage to various interval sizes or do they act as some kind of musical score? Or do they in fact actually play a type of "music" by directing energies to vibrate at specific frequencies out into space? Like the "square" craters, they cover a range of 5 octaves but are about half the physical size of their counterparts. The presence of 2 types of craters encoding musical "notes" seems almost analogous to having 2 different keyboards for a musical instrument such as is found in many of our church organs.

If we look back at the magnitudes of the Pentagon and Pentagram Pyramids in the light of how pentagon craters are sized, we would find that the Pentagram Pyramid with a radius of 0.25 latitude degrees would fit in the chromatic scale at the P0 interval for octave -2 in Table 3.2. Since the Pentagon Pyramid is about 1.44 times the size of the Pentagram Pyramid, its size would make it a variant of the diminished 5th interval, which is an interesting way to refer to the number 5 assuming the Martians used a system similar to ours for identifying music intervals. The exact interval cannot be determined yet since the measurements of the Pentagon Pyramid are not accurate enough to permit this. Plausible intervals include 36:25 and 13:9.

There is an uneven distribution of pentagon craters across longitude and latitude. They are absent from the longitudes occupied by the big mountains. This may be due to a desire to restrict different spiritual messages to separate regions. The mountains emphasize the sacred geometry of the Martian body shape. Perhaps the pentagon craters are supposed to emphasize sacred music. It is more difficult to come up with an explanation for why there is a concentration of pentagon craters at latitudes which are distant from the equator. Perhaps these latitudes were more often used for flight paths.

The finding that one or more elements of the pentagons were located at coordinates which had the value of an integer, integer and one-half, or a multiple of π, π^2, φ, φ^2, e, e^2, $\sqrt{5}$, $\sqrt{3}$ or $\sqrt{2}$ suggests that they were used to honour certain numerical values. Some of the integers refer to the pentagram and other aspects of sacred geometry. Perhaps some form of numerology was at play for integer values which do not seem to relate to sacred geometry. All the values using irrational numbers, however, most likely do refer to sacred geometry. The integer and integer plus 1/2 coordinate values do not seem to identify latitudes or longitudes in a systematic way so it is unlikely that the pentagon craters were used as a coordinate grid map.

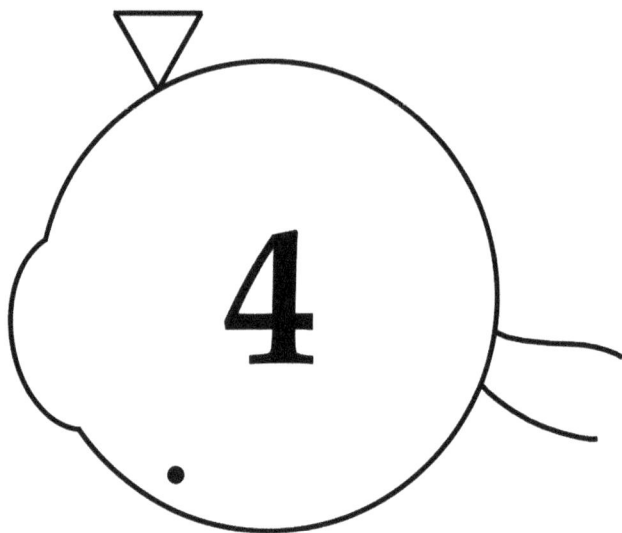

Hexagon Craters

In Chapter 1 there was brief mention of 2 hexagon-shaped craters which had stimulated me to think in terms of different degree sizes since the width of one in longitude degrees was approximately equal to 1.125 regular degrees and that of the other was close to 1.25 degrees. These craters could be fit with a regular hexagon which is a 6-sided polygon in which the sides are all equal and the internal angles are all 120° in size. When you examine the MOLA maps of Mars closely, it becomes obvious that a substantial number of the craters have the basic shape of a regular hexagon. Linear segments of the perimeters of these craters follow enough of the outline of a regular hexagon to enable a fit to be made, but like the "square" and "pentagon" craters, the perimeter never totally aligns with it. The sides of the fitted hexagon make contact with the crater perimeter in regions other than just the centres of the sides, thus negating the possibility that the crater is simply a circle which fits inside the hexagon. I decided that it was important to accurately measure a large group of these craters in order to determine if their sizes, locations and orientation would provide some clue as to what their role might have been. Since it is much easier to work with the Mercator projection maps, I restricted myself to the region of the planet between 30° S and 30° N.

The first crater which I will show you is one in which the hexagonal shape of its perimeter is quite well defined (Fig. 4.1). As the crater is unnamed, I called it nn68 in my numbering system. The centre of the

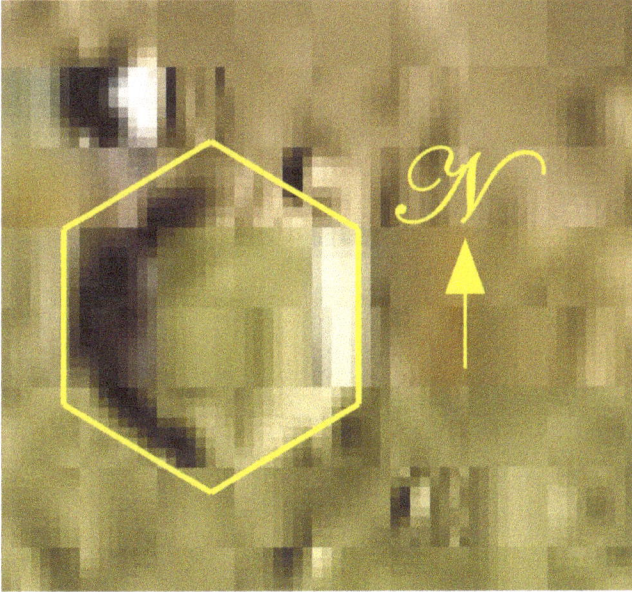

Fig. 4.1: *Hexagon-shaped crater nn68 at 41.8163° E 28.5228° S. It has a side-to-side width of 0.3000 latitude degrees, and 2 vertices are aligned in a north-south direction. Its northern vertex is 18.0000°° S of the Arsia Mons Prime Latitude. The number 18 is half the size of a star point angle in a pentagram in regular degrees. USGS Astrogeology.*

hexagon which was fitted to this crater sits at 41.8163° E 28.5228° S. It is relatively small with a side-to-side width of 0.3000 latitude degrees or about 17.78 km. When a regular hexagon is fitted to its perimeter, most of the crater perimeter follows the outline of the hexagon. Notice that one of the hexagon vertices points due north and another due south. When the coordinates of the vertices are examined, it is found that the latitude of the northern vertex is 18.0000°° S (AMPL). The average longitude of the east side of the fitted hexagon is also very interesting. It is $99\varphi^2$ = 259.1843° W, $88\varphi^2$ = 230.3870°° W, and 207.3483°°° W which is close to 66π = 207.3451°°° W (all with reference to the Sharonov Tower PM). Note that all of these values are evenly divisible by 11.

The next hexagon-shaped crater (nn71) is shown in Fig. 4.2 and it is also relatively small, having a side-to-side width of 0.2667 latitude degrees or about 15.81 km. The centre of the fitted hexagon is located at 95.0627° E and 17.1872° N. This time the fitted hexagon has a vertex which points due east and another which points due west. The north side of the hexagon has a latitude of $10\sqrt{3}$ = 17.3205° N and $8\sqrt{3}$ = 13.8564°°° N. The south side of the hexagon has a latitude (with respect to the Arsia Mons Prime Latitude) very close to $10\sqrt{5}$°° N (22.3587°° vs. 22.3608°°) and $9\sqrt{5}$°°° N (20.1228°°° vs. 20.1246°°°). The average longitude of the northeastern and southeastern vertices is $30\sqrt{3}$ = 51.9615° W and $24\sqrt{3}$ = 41.5692°°° W (Crater Edge PM). This longitude is also very close to $30\sqrt{2}$°°° W of the Dagger Midline PM (42.4233°°° vs. 42.4264°°°). Hence, with the coordinate values of its vertices, this crater is especially focused

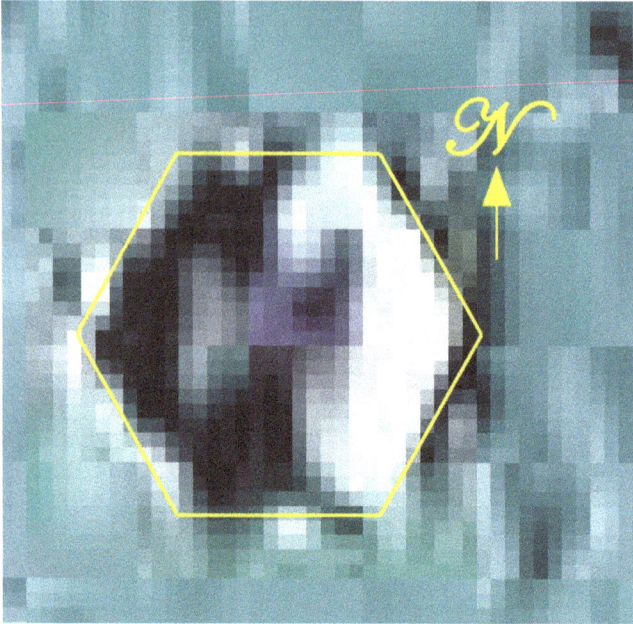

Fig. 4.2: *Hexagon-shaped crater nn71 fitted with a hexagon having a centre located at 95.0627° E 17.1872° N. The hexagon has a side-to-side width of 0.2667 latitude degrees, and 2 vertices are aligned in a east-west direction. The north side of the hexagon lies at $10\sqrt{3}°$ N. USGS Astrogeology.*

on $\sqrt{3}$ but it also celebrates the sacred irrational numbers of $\sqrt{2}$ and $\sqrt{5}$.

Fitting a Crater to Concentric Hexagons

Some craters can be fit to more than 1 hexagon. One way in which this is done is to fit a crater to hexagons which are almost concentric. In Fig. 4.3, I have fitted a large unnamed crater (nn81 in my numbering scheme) to a hexagon centred at 346.0162° E 10.8170° N whose side-to-side width is equal to 1.5 degrees of latitude or about 88.91 km. The hexagon has one vertex pointing due north and another pointing due south. I chose the location of the fitted hexagon on the basis of features of the crater perimeter as well as values of the coordinates of some of its vertices. The west side lines up with the crater perimeter in the middle of the side and near the northwest vertex. The northeast side aligns with the northern edges of a dark region near the north vertex and a white section of the wall of the crater further to the southeast. The east side aligns with the crater perimeter in a region near the northeast vertex and in a region just south of the middle of the east side. The southeast and southwest sides of the hexagon align with structures within the crater wall near the south vertex. The latitude of the northeast and northwest vertices is 11.2500° N, 10.0000°° N and 9.0000°°° N. The longitude of the north and south vertices and of the hexagon centre is 176.7500°° E (Elysium Mons PM).

A second hexagon can be fit to this crater by increasing the side-to-side

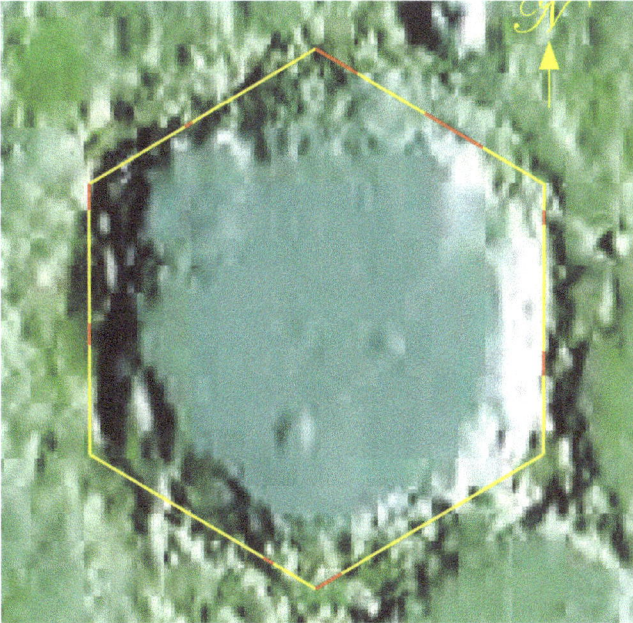

Fig. 4.3: *Large crater (nn81) fitted to a hexagon with 2 of its vertices pointing north-south. The hexagon is 1.5 latitude degrees from side-to-side and fits many sections of the crater perimeter. Alignments to various aspects of the crater are indicated by short red lines superimposed upon the hexagon. The latitude of the northeast and northwest vertices is 11.2500° N, 10.0000°° N and 9.0000°°° N. USGS Astrogeology.*

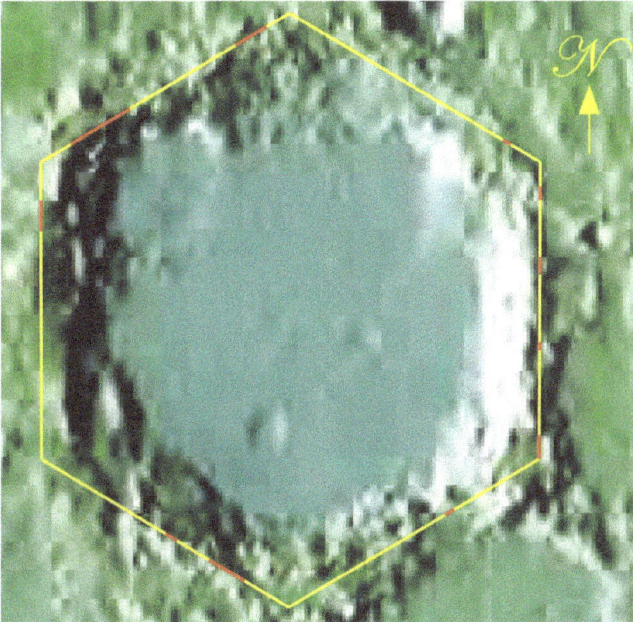

Fig. 4.4: *The hexagon in Fig. 4.3 moved to a slightly different centre and enlarged to 1.6 latitude degrees for its side-to-side dimension. The northwest side aligns with a section of the crater perimeter near the northwest vertex as does part of the west side. Alignments of the hexagon to various aspects of the crater are indicated by short red lines. The south vertex of the hexagon is at 18.0000° N and 16.0000°° N (both AMPL). USGS Astrogeology.*

width to 1.6 degrees of latitude and moving the latitude of the centre slightly from 10.8170° N to 10.8242° N (Fig. 4.4). This larger hexagon fits the crater perimeter extremely well on the northwest side near the northwest vertex. The west side shows a good alignment to a linear portion of the crater perimeter near the same vertex. The northern part of the east side touches the crater perimeter for a short length near the

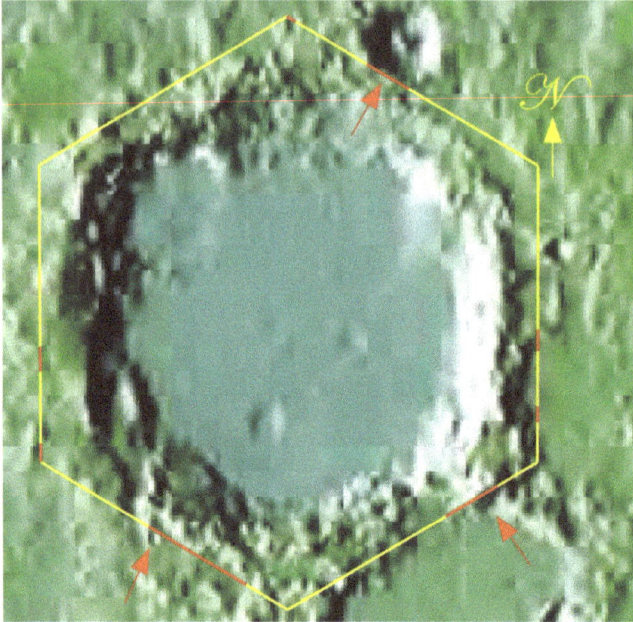

Fig. 4.5: *The hexagon in Fig. 4.3 moved to a slightly different centre and enlarged to 1.8 latitude degrees for its side-to-side dimension. The hexagon lines up with the furthest extension of the crater perimeter on the southeast side. It also aligns to the edges of 3 auxiliary craters (red arrows). USGS Astrogeology.*

northeast vertex, and other sections of the side align with structures just beyond the crater perimeter. The southwest side aligns with the crater perimeter for much of the region between the middle of the side and the southern vertex. The latitude of the southern vertex is 18.0000° and 16.0000°° north of the Arsia Mons Prime Latitude (AMPL).

If the hexagon size is increased to a side-to-side width of 1.8 latitude degrees and the centre shifted slightly to 346.0211° E 10.8358° N you obtain the fit shown in Fig. 4.5. The southeast side lines up with a section of the crater perimeter and the southwest side with a long portion of the crater apron. The northeast side aligns with the linear edge of a sizeable crater to the northeast of the hexagon-shaped crater. This linear edge seems to be the limit of the nn81 crater's northeast perimeter. The southeast and southwest sides also touch the edges of small craters. Thus 3 auxiliary craters (indicated by red arrows in Fig. 4.5) help to define the borders of the largest hexagon. The northern vertex of the hexagon is at 9.5000°°° N. The average longitude of the west side is at 197.0000° E (Dagger Peak PM), 198.0000° E (Crater Edge PM), 98.0000° E (Pavonis Mons PM) and 44.0000° E (Sharonov Triangle PM).

I found another example of a hexagon-shaped crater (unnamed crater, E604 in my numbering system) which could be fit to almost concentric hexagons. This crater is also quite large having a diameter of about 67 km. A hexagon (centred at 93.7974° E 8.7000° S) with a side-to-side width of 1.125 latitude degrees fits the crater perimeter extremely well on the

Fig. 4.6: *Crater E604 fitted to a hexagon (centre coordinates 93.7974° E 8.7000° S) with 2 vertices oriented in the north-south direction. A fitted hexagon with a side-to-side width of 1.125 latitude degrees aligns very well to the crater perimeter on the northwest and southwest sides. The southern vertex is 1.0000°°° south of the Arsia Mons Prime Latitude. USGS Astrogeology.*

northwest and southwest sides (Fig. 4.6). The west side aligns with a short linear region of the perimeter near the northwest vertex, and with the edge of a structure within the crater wall. The east side aligns with the crater perimeter near the southeast vertex and just north of the middle of the side. It also aligns with the edges of 2 depressions within the crater wall. The northeast side aligns with a short section of the crater perimeter near the north vertex. The southern vertex of the hexagon is 1.2500° and 1.0000°°° south of the Arsia Mons Prime Latitude. The north and south vertices and crater centre are at 43.5000°°° W (Dagger Midline PM).

If the hexagon size is increased to 1.20 latitude degrees and the centre shifted slightly to 93.7794° E 8.6928° S it will then fit the southeast crater perimeter over a very long distance (Fig. 4.7). Its west side aligns with a short section of the crater perimeter and with the edge of a light coloured structure on the lip of the crater. A short section of the crater perimeter also aligns with the northeast side of the hexagon. The north vertex of the hexagon sits at 8.0000° S and the average longitude of the west side of the hexagon is at 54.0000° W and 48.0000°° W (Elysium Mons PM). The value of 54 is 1/2 the size of the internal angles of a pentagon or the angles between the star points of a pentagram.

By increasing the side-to-side width to 1.25 latitude degrees or about 74.09 km and shifting the centre slightly to 93.7901° E 8.7031° S, the hexagon now fits the eastern limits of the crater perimeter and the outermost linear section of the western perimeter (Fig. 4.8). The southeast

Fig. 4.7: *The hexagon in Fig. 4.6 moved to a slightly different centre and enlarged to 1.200 latitude degrees for its side-to-side dimension. The southeast side has a major alignment with the crater perimeter. The west side aligns with a small section of the crater perimeter and with the edge of a light coloured structure. The north vertex of the hexagon is at 8.0000° S. USGS Astro-geology.*

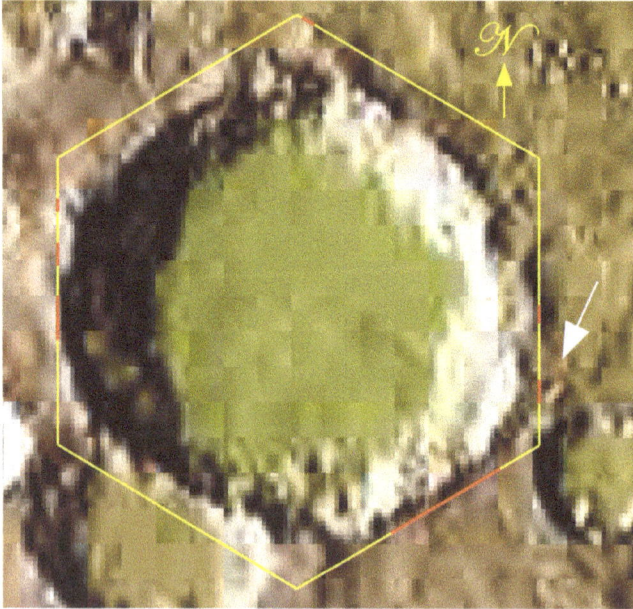

Fig. 4.8: *The hexagon in Fig. 4.6 moved to a slightly different centre and enlarged to 1.250 latitude degrees for its side-to-side dimension. The southeast side of the hexagon aligns with the edge of the perimeter apron. The west and east sides align with short sections of the crater perimeter. As well, the east side aligns with the edge of a small auxiliary crater (white arrow). The south vertex is at $3\pi°$ S. USGS Astrogeology.*

side is aligned to a long linear portion of the outer edge of the apron of the crater perimeter. The northeast side aligns to a very small region of the crater perimeter near the north vertex. The east side aligns with the edge of a tiny auxiliary crater near the southeast vertex (white arrow). The south vertex has the remarkable latitude of $3\pi = 9.4248°$ S The average longitude of the east side is 53.7500° W and 43.0000°°° W (Dagger Midline PM).

Fitting a Crater to Multiple Hexagon Rotations

A second way in which a hexagon-shaped crater can be fit to more than 1 hexagon is by rotating the hexagon and fitting it to different alignment sites. The unnamed crater E714 can be fit to a hexagon whose side-to-side width is 0.75 latitude degrees (44.46 km). The hexagon (upper left picture of Fig. 4.9) has vertices pointing north and south. The crater perimeter touches all sides except the southwest side. The west and east sides of the hexagon align with short and long linear stretches of the crater perimeter. The hexagon centre is located at 8.3106° E 15.0000° S. The latitude can also be expressed as 12.0000°°° S. The importance of the number 12 has been encountered before in the Pentagon Pyramid and in the Martian Meter. The average longitude of the hexagon's west side is 139.25° W (Elysium Mons PM), 140.25° W (Dagger Midline PM) and 293.25° W (Sharonov Tower PM) which gives credence to a 1440 degree system.

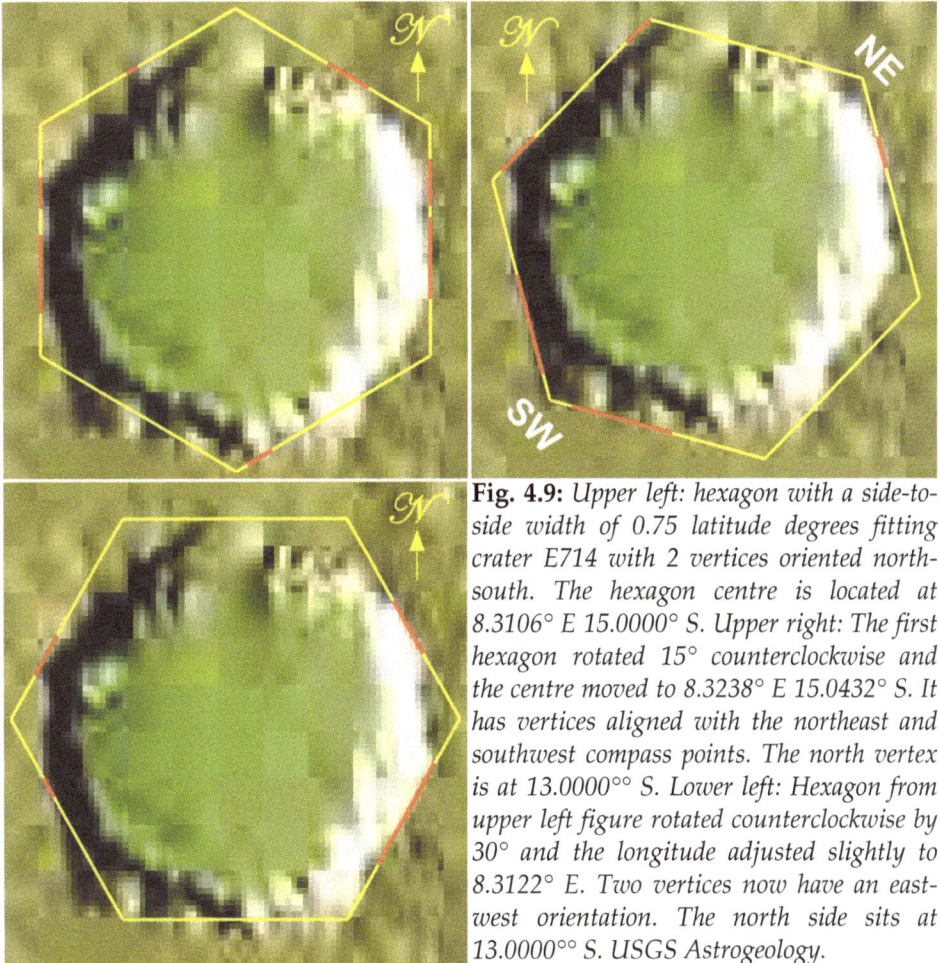

Fig. 4.9: *Upper left: hexagon with a side-to-side width of 0.75 latitude degrees fitting crater E714 with 2 vertices oriented north-south. The hexagon centre is located at 8.3106° E 15.0000° S. Upper right: The first hexagon rotated 15° counterclockwise and the centre moved to 8.3238° E 15.0432° S. It has vertices aligned with the northeast and southwest compass points. The north vertex is at 13.0000°° S. Lower left: Hexagon from upper left figure rotated counterclockwise by 30° and the longitude adjusted slightly to 8.3122° E. Two vertices now have an east-west orientation. The north side sits at 13.0000°° S. USGS Astrogeology.*

If the hexagon is rotated by 15 degrees in the counterclockwise direction and the centre moved to 8.3238° E 15.0432° S, it fits the crater perimeter in different locations (Fig. 4.9, upper right). The west and southwest sides of the hexagon are aligned to long linear regions of the crater perimeter to the west and south. The east and northwest sides also touch short linear regions of the crater perimeter. Two of the hexagon vertices now correspond to the northeast and the southwest compass points. The latitude of the north vertex is 13.0000°° S and that of the southeast vertex is 12.1243°°° S with a numerical value almost exactly equal to $7\sqrt{3} = 12.1244$. The longitude of the north vertex is 191.1174°°° W (Pavonis Mons PM) with a numerical value close to $73\varphi^2 = 191.1165$.

Rotating the original hexagon by 30 degrees in the counterclockwise direction puts a vertex facing due east and another facing due west (Fig. 4.9, lower left). I moved the centre of this hexagon slightly to 8.3122° E 15.0000° S. The new hexagon aligns to short linear segments of the crater perimeter on its west sides and to long linear segments on its east sides. The latitude of the east vertex, the west vertex and the hexagon centre is at 15.0000° S and 12.0000°°° S. The latitude of the north side of the hexagon is 13.0000°° S, the same as for the latitude of the north vertex in the previous rotation (this happens because the centre of the previous rotation is located further south). The northeast and southeast vertices sit at an average longitude of $81\varphi^2 = 212.0608°°$ W (Pavonis Mons PM).

One last note on this crater: the latitude of the centres for the original hexagon and its 30° rotation is 15° S. This is the exact number of degrees per rotation. This may be a clever placement to emphasize the number 15 which has as factors the 2 very important numbers of 3 and 5 which are also Fibonacci numbers. Using 15 degree rotations ensures that 2 vertices of each hexagon correspond to points on a compass divided into 8 major directions provided that you start with a hexagon already aligned to a pair of opposite compass directions. However, the 15° rotations of its hexagons are not unique to this crater as we shall see next.

Another crater that I found to fit rotated hexagons is an unnamed crater (nn75) which is fit very well with a hexagon having a side-to-side width of 0.625 latitude degrees and centred at 172.1724° E 28.0996° S (Fig. 4.10, upper left). Four sides of the hexagon run along lengthy linear segments of the crater perimeter. There are, however, a couple of curious notches in the crater perimeter, one extending northwards beyond the northwest side and the other extending southwards beyond the southwest side. The latter notch is used to fit the hexagon in the lower right picture of Fig. 4.10, but I did not discover a function for the northern notch. The centre of the upper left hexagon has amazing coordinates. When referenced to the Arsia Mons Prime Latitude it is 20.0000° S and

Fig. 4.10: *Upper left: a hexagon oriented in the north-south direction with a side-to-side width of 0.625 latitude degrees fits the nn75 crater at 172.1724° E 28.0996° S. The hexagon centre is 20.0000° S (AMPL) and 16.0000°°° S (AMPL) and 25.0000° E and 20.0000°°° E (Elysium Mons PM). Upper right: hexagon rotated 15° counterclockwise and centre latitude changed to 28.1250° S. This hexagon fits the crater perimeter very well on the northwest side. The latitude of the centre is 25.0000°° S and 22.5000°°° S. Lower left: first hexagon rotated 30° counterclockwise and centred at 172.2178° E 28.1250° S. This hexagon is oriented east-west. The southeast side aligns to the edge of the dark apron extending southwards from the southern crater perimeter. The west vertex is at 21.0100°° E (Dagger Midline PM). Lower right: first hexagon rotated 45° counterclockwise and centred the same as for the 15° rotation. This hexagon fits the crater perimeter especially well on the southeast and southwest sides. USGS Astrogeology.*

16.0000°°° S. The centre's longitude and that for the north and south vertices is 25.0000° E and 20.0000°°° E (Elysium Mons PM), 24.0000° E

(Dagger Midline PM), 75.0000° W and 60.0000°°° W (Pavonis Caldera PM), and 129.0000° W (Sharonov Tower PM). Note that there is a match in longitude and latitude of 20°°° E (EMPM) 20° S (AMPL) which creates a Rosetta Stone for sacred degrees analogous to the big degree Rosetta Stone of 12°° E 12° N (DPPM) for the Pentagon Pyramid east-west midpoint. It also validates the Arsia Mons Prime Latitude and the Elysium Mons PM.

When the hexagon is rotated by 15 degrees in the counterclockwise direction and the latitude of the centre moved slightly to 28.1250° S while keeping the longitude the same, the northwest side of the hexagon fits the perimeter very well (Fig,. 4.10, upper right). Since none of the other sides show alignments, the fit of this hexagon depends on the size obtained by the previous rotation, i.e., not enough information is present to define it on its own (this is also true for the next rotation below). The latitude of the centre of the hexagon is 25.0000°° S and 22.5000°°° S which creates another match in latitude and longitude (same as for previous hexagon's centre), i.e., 25° E 25°° S (EMPM) analogous to the 12°° E 12° N (DPPM) coordinates for the Pentagon Pyramid east-west midpoint. This time, however, the coordinate in big degrees is for latitude as opposed to longitude for the Pentagon Pyramid east-west midpoint. One thing more to mention is that this hexagon has one vertex aligned with the NE compass point and another vertex aligned with the SW compass point.

Rotating by another 15 degrees in the counterclockwise direction and changing the coordinates to 172.2178° E 28.1250° S gives a hexagon with a pair of vertices that point due east and due west (Fig. 4.10, lower left). Here we see a fit of the southeast side to the edge of the dark region of the crater apron extending southward. The south side fits a short stretch of the crater perimeter near the southwest vertex. This hexagon also has interesting coordinates. The latitude of the hexagon centre and the east and west vertices is 25.0000°° S and 22.5000°°° S. The north side sits at 22.2500°°° S and the south side at 22.7500°°° S which gives credence to a 1152 degree system. The northeast and southeast vertices are at an average longitude of 103.0000°°° W (Sharonov Tower PM). The west vertex is at 21.0100°° E (Dagger Midline PM).

Rotating the hexagon a final 15 degrees in the counterclockwise direction and putting the centre at the same coordinates as the hexagon for the first 15 degree rotation (Fig. 4.10, upper right) produces a hexagon (Fig. 4.10, lower right) whose southwest side fits the linear portion of the crater perimeter forming the southern notch as mentioned above. The northeast side also fits a linear portion of the crater perimeter and the southeast side touches the crater perimeter at several locations over a long distance. The longitude of this hexagon's centre with reference to the

Table 4.1: *Hexagon-shaped craters between 30°S to 30°N. NS = north - south, EW = east - west, NESW = northeast - southwest and SENW = southeast - northwest. Craters are listed in the order of their side-to-side width in units of latitude degrees.*

Crater Name	Orientation	Width (°)	Longitude (°E)	Latitude (°N)	Crater Name	Orientation	Width (°)	Longitude (°E)	Latitude (°N)
nn73	NS	0.2083	121.0608	-21.0855	nn61	EW	0.5333	17.9906	14.5855
nn71	EW	0.2667	95.0627	17.1872	nn74	NS	0.5333	159.8552	-14.4456
E875	NS	0.2667	44.4224	2.1174	nn65	EW	0.6000	34.0486	-1.6180
E952	NS	0.2667	49.6252	26.3342	nn75	NS	0.6250	172.1724	-28.0996
E945	EW	0.2813	50.0145	25.8224	nn75	NESW	0.6250	172.1724	-28.1250
nn62	NS	0.3000	19.5184	13.1888	nn75	EW	0.6250	172.2178	-28.1250
nn68	NS	0.3000	41.8163	-28.5228	nn75	SENW	0.6250	172.1724	-28.1250
nn79	NS	0.3000	338.3547	-28.2500	E700	NS	0.7111	10.3550	20.6359
E868	NS	0.3125	43.6047	5.2805	E1004	EW	0.7111	71.8195	17.0562
E735	NS	0.3333	18.9793	23.6250	E714	NS	0.7500	8.3106	-15.0000
nn63	NS	0.3333	23.3603	27.5000	E714	NESW	0.7500	8.3238	-15.0432
E938	EW	0.3333	118.5644	-9.6129	E714	EW	0.7500	8.3122	-15.0000
E603	NS	0.3556	99.8925	17.2614	E707	NS	0.8333	12.0460	19.9042
nn78	NS	0.3556	210.3301	-14.7676	nn77	NS	0.8889	212.9269	-10.5864
nn64	NS	0.4000	33.9042	4.2691	nn70	EW	0.9375	43.0934	20.5313
H938	EW	0.4000	47.3152	9.4988	E604	NS	1.1250	93.7974	-8.7000
E1011	EW	0.4000	74.7437	11.8000	E604	NS	1.200	93.7794	-8.6928
nn72	NS	0.4000	109.4849	-28.2405	Masked	NS	1.200	139.6734	-17.5000
E749	NS	0.4167	24.0571	26.1803	E604	NS	1.2500	93.7901	-8.7031
E959	NS	0.4500	47.6414	26.8512	E596	NS	1.2500	30.8066	2.1811
nn66	EW	0.4688	34.7763	-2.4704	nn81	NS	1.5000	346.0162	10.8170
E601	NS	0.5000	17.1724	17.7113	nn81	NS	1.6000	346.0162	10.8242
E600	NS	0.5333	19.0097	17.7631	nn81	NS	1.8000	346.0211	10.8358

Martian prime meridians, and its latitude, are the same as for the first 15 degree rotation. Notice that one vertex aligns to the NW compass point and another to the SE compass point.

Overall Picture of Hexagon-Shaped Craters

I found a total of 37 hexagon-shaped craters between 30° S and 30° N. Two of these could be fit to concentric hexagons and another two could be fit to hexagons differing in rotational angle as mentioned above. This created a total set of 46 hexagons (Table 4.1). Out of the 46 hexagons, 31 had vertices pointing north and south, 12 had vertices pointing east and west, 2 had vertices pointing to the northeast and southwest compass points, and 1 had vertices pointing to the southeast and northwest compass points. Assuming a resolution of ±1 degree of rotation, the probability of getting 31 out of 46 hexagons with vertices pointing north and south under random conditions is less than 1 in a trillion trillion which is extremely good evidence of artificiality.

Of the 37 unique craters, none were found in the region from Olympus Mons to the eastern edge of the Valles Marineris. The vast majority (22 or 59%) were found between 0° E - 60° E which is on the opposite side of the planet from Olympus Mons. They were also unevenly distributed across latitude with 18 occurring in the range of 10° N to 30° N as opposed to 19 occurring in the much wider range of 30° S to 10° N.

Size of the fitted hexagons varied from 0.2083 to 1.8 latitude degrees for the side-to-side distance which seems to be the parameter that was used by the architects for sizing hexagons. Like the square and pentagon craters, all sizes can be expressed in terms of the ratios of integers which relate to the interval sizes of the 12 note chromatic scale. In Table 4.2, a hexagon width of one degree of latitude is set to a perfect unison interval and all other sizes are related to that size either in the same octave or in octaves below this value. In this way, the smallest crater has the size of a major 6th note in the lowest octave and the largest note has a value of a minor 7th three octaves above this. All notes of the chromatic scale are present in the hexagon crater dataset. If we exclude repeated sizes for rotations of hexagons, the most frequently appearing notes are a minor 2nd (6 times), a minor 3rd (6 times) and a minor 6th (5 times). The tritone interval which did not occur in the square-shaped or pentagon-shaped craters occurs 4 times in the hexagon-shaped craters. The integer ratio for the tritone interval was found to be 64:45 rather than the slightly lower pitched integer ratio of 45:32. The perfect unison (P0) interval occurs only once as opposed to 6 times for the square-shaped craters and 5 times for the pentagon-shaped craters.

All of the hexagons had vertices and/or a centre which were at meaningful coordinate values (Table 4.3). Most of these values were either pure integers or involved the use of one of the 6 basic irrational numbers (π, φ, e, $\sqrt{2}$, $\sqrt{3}$, or $\sqrt{5}$). One of the longitudes was an integer plus 1/4 value, another was an integer plus 1/2 value, and 3 were integer plus 3/4 values. There is a very strong reference to the pentagram in Table 4.3 with 17 longitudes and 25 latitudes using numbers which could relate to this geometric figure (φ, 10, 25, 36, 18, 9, 54, 27). The most often used prime meridian was the Dagger Midline PM (16 instances), followed by the Elysium Mons PM (9 instances), the Crater Edge PM (7 instances) and the Dagger Peak PM (6 instances). The equator was most often used as a

Table 4.2: *Size of hexagon-shaped craters in terms of notes in the chromatic scale. M=major, m=minor, P=perfect and tt=tritone (64:45).*

Crater Name	Width (°)	Inter-val Ratio	Inter-val Name	Octave No.	Crater Name	Width (°)	Inter-val Ratio	Inter-val Name	Octave No.
nn73	0.2083	5/3	M6	-3	nn61	0.5333	16/15	m2	-1
nn71	0.2667	16/15	m2	-2	nn74	0.5333	16/15	m2	-1
E875	0.2667	16/15	m2	-2	nn65	0.6000	6/5	m3	-1
E952	0.2667	16/15	m2	-2	nn75	0.6250	5/4	M3	-1
E945	0.2813	9/8	M2	-2	nn75	0.6250	5/4	M3	-1
nn62	0.3000	6/5	m3	-2	nn75	0.6250	5/4	M3	-1
nn68	0.3000	6/5	m3	-2	nn75	0.6250	5/4	M3	-1
nn79	0.3000	6/5	m3	-2	E700	0.7111	64/45	tt	-1
E868	0.3125	5/4	M3	-2	E1004	0.7111	64/45	tt	-1
E735	0.3333	4/3	P4	-2	E714	0.7500	3/2	P5	-1
nn63	0.3333	4/3	P4	-2	E714	0.7500	3/2	P5	-1
E938	0.3333	4/3	P4	-2	E714	0.7500	3/2	P5	-1
E603	0.3556	64/45	tt	-2	E707	0.8333	5/3	M6	-1
nn78	0.3556	64/45	tt	-2	nn77	0.8889	16/9	m7	-1
nn64	0.4000	8/5	m6	-2	nn70	0.9375	15/8	M7	-1
H938	0.4000	8/5	m6	-2	E604	1.1250	9/8	M2	0
E1011	0.4000	8/5	m6	-2	E604	1.200	6/5	m3	0
nn72	0.4000	8/5	m6	-2	Masked	1.200	6/5	m3	0
E749	0.4167	5/3	M6	-2	E604	1.2500	5/4	M3	0
E959	0.4500	9/5	m7	-2	E596	1.2500	5/4	M3	0
nn66	0.4688	15/8	M7	-2	nn81	1.5000	3/2	P5	0
E601	0.5000	1/1	P0	-1	nn81	1.6000	8/5	m6	0
E600	0.5333	16/15	m2	-1	nn81	1.8000	9/5	m7	0

latitude reference (34 instances vs. 12 instances of the use of the Arsia Mons Prime Latitude). For longitudes, the pair of eastern vertices (northeast and southeast) were most often used (16 instances vs. 7 instances for the western pair) and the east-most vertex was used 5 times vs. only once for the west-most vertex. For latitudes, the pair of northern vertices (northeast and northwest) were used more often (10 instances) than the pair of southern vertices (4 instances) although the south-most vertex (10 instances) and the north-most vertex (9 instances) were used with about the same frequency.

Crater	Element	Longitude	P.M.	Element	Latitude	P.L.
nn73	ne, se	27° W	DMPM	s	6π° S	
nn71	ne, se	30√3° W	CEPM	nw, ne	10√3° N	
E875	n, s, c	83°°° W	DMPM	s	(φ/2)°°° N	
E952	sw, nw	(10π²)° W	DMPM	s	10(φ²)° N	
E945	ne, se	98° W	DPPM	nw, ne	atan(φ/π)°°° N	AMPL
nn62	ne, se	102°°° W	EMPM	nw, ne	19°° N	AMPL
nn68	ne, se	99(φ²)° W	SToPM	n	18°° S	AMPL
nn79	n, s, c	73°°° E	CEPM	c	28.25° S	
E868	n, s, c	92°° W	CEPM	n	3φ° N	
E735	sw, nw	115°° W	DMPM	c	21°° N	
nn63	sw, nw	100°°° W	DMPM	c	22°°° N	
E938	n	18√2°° W	CEPM	sw, se	(e/φ)° S	AMPL
E603	ne, se	178.75°° W	SToPM	s	9√5°°° N	AMPL
nn78	n, s, c	44√2° W	DPPM	n	9φ° S	
nn64	ne, se	114° W	DPPM	n	4°° N	
H938	e	36√5°°° W	DPPM	nw, ne	11φ° N	AMPL
E1011	e	65°° W	DPPM	nw, ne	12° N	
nn72	n, s, c	33.5°° W	EMPM	nw, ne	25°° S	
E749	n, s, c	47(φ²)° W	CEPM	c	10(φ²)° N	
E959	n, s, c	32π° W	DMPM	nw, ne	asin(1/e)°°° N	
nn66	e	54π°°° W	PCPM	nw, ne	√5° S	
E601	n, s, c	130° W	EMPM	n	18° N	
E600	e	80φ° W	DMPM	sw, se	7√5°° N	
nn61	w	116°° W	DMPM	sw, se	9√2°° N	
nn74	c	7φ°° E	CEPM	n	10√2° S	
nn65	ne, se	29π°°° W	STrPM	w, e, c	φ° S	
nn75a	c	25° E	EMPM	c	20° S	AMPL
nn75b	c	25° E	EMPM	c	25°° S	
nn75c	ne, se	103°°° W	SToPM	w, e, c	25°° S	
nn75d	c	25° E	EMPM	c	25°° S	
E700	ne, se	35π°°° W	DMPM	s	10φ° N	
E1004	e	29(φ²)° W	DMPM	w, e, c	10√5°° N	AMPL
E714a	sw, nw	140.25° W	DMPM	c	15° S	
E714b	se	71√3°° W	CEPM	n	13°° S	
E714c	ne, se	81(φ²)° W	PMPM	w, e, c	12°°° S	
E707	ne, se	116φ°°° W	PMPM	s	9e°° N	AMPL
nn77	sw, nw	33√3°° E	DMPM	s	3° S	AMPL
nn70	ne, se	40(φ²)° W	DPPM	nw, ne	21° N	
E604a	n, s, c	43.5°°° W	DMPM	s	1°°° S	AMPL
E604b	sw, nw	54° W	EMPM	n	8° S	
Masked	ne, se	7°° W	DMPM	c	14°°° S	
E604c	ne, se	43°°° W	DMPM	s	3π° S	
E596	ne, se	66√2°°° W	DPPM	sw, se	φ°° N	
nn81a	n, s, c	176.75°° E	EMPM	nw, ne	10°° N	
nn81b	n, s, c	176.75°° E	EMPM	s	18° N	AMPL
nn81c	sw, nw	44° E	STrPM	n	9.5°°° N	

Table 4.3: *Meaningful coordinates for hexagons fitting hexagon-shaped craters. Hexagon elements include the hexagon centre (c) and the 6 vertices (n = north, s = south, e = east, w = west, nw = northwest, ne = northeast, sw = southwest, se = southeast). P.M. = Prime Meridian and P.L. = Prime Latitude. Note that coordinates are in terms of regular degrees (°), big degrees (°°) or sacred degrees (°°°).*

Summary and Conclusions

The hexagon-shaped crater is another category of crater which is sized according to chromatic scale interval values. Since these craters all have vertices which point either to the 4 cardinal directions of the compass or to the directions halfway between the cardinal directions, it is possible that the craters served as bidirectional compasses for overhead spacecraft. For those craters which I found to serve as templates for different rotations of the same size of hexagon, the compass would be more complete although it would be rather asymmetrical since the centre of the hexagon shifts between these hexagons.

Another very good possibility is that the fitted hexagons were used to mark certain coordinate values that were considered to be important numbers either from a spiritual or aesthetic perspective. There was no evidence of a systematic marking of coordinates which would provide a grid map for overhead spacecraft. The use of music intervals for sizes also suggests a spiritual or aesthetic purpose behind the craters. Their absence from the area of the giant mountains is likely to be due to a desire to not interfere with the symbolism of these other Martian feats of architecture. Since the hexagon-shaped craters were also absent from the area of the Valles Marineris, it makes one wonder if this region had some spiritual significance of its own.

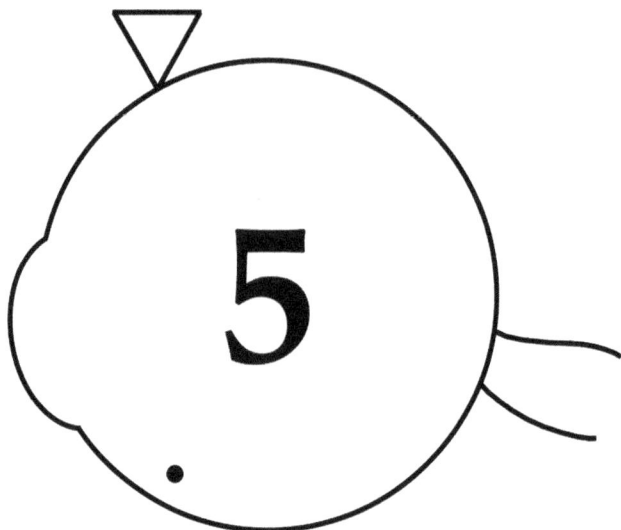

Octagon Craters

W hile scanning the map for hexagon-shaped craters, I discovered 2 craters which had the shape of a regular octagon i.e., a polygon with 8 equal sides and 8 equal internal angles of 135°. These unnamed craters were located on the eastern edge of the largest group of hexagon craters and were close to the equator. They both have the same orientation, with 2 parallel sides running east-west and another 2 parallel sides running north-south.

The first crater that I will present (Fig. 5.1) is located just south of the equator at 71.8547° E 1.5549° S. It can be well fit to an octagon with a side length of 0.2813 latitude degrees. The octagon vertices situated at the north ends of the east and west sides have a latitude of $\sqrt{2}$ = 1.4142° S. They are also at $8(\sqrt{2}/10)$ = 1.1314°°° S expressed in sacred degrees which is a remarkable reference to the 8 sides of the octagon. The $\sqrt{2}$ latitude value is notable in itself since $\sqrt{2}$ is found extensively in the structure of the octagon when the octagon is oriented in the same way as in Fig. 5.1. If we set s = the length in latitude degrees of one of the octagon sides, the latitude (°S) of the northern side of the octagon is equal to the latitude (°S) of the centre - s/2 - s/$\sqrt{2}$. Likewise, the longitude (°E) of the eastern side is the longitude (°E) of the centre + s/2 + s/$\sqrt{2}$. This happens because the bearing angle of the sides not aligned north-south or east-west is 45° in either the clockwise or counterclockwise direction. This makes the distance covered by these sides in either the north-south direction or the

Fig. 5.1: *Octagonal-shaped crater at 71.8547° E 1.5549° S. The side length of the fitted octagon is 0.2813 latitude degrees. The octagon touches the crater perimeter on all sides. The northern vertices of the east and west sides have a latitude of √2° S. The centre of the octagon is 61.0000°°° W (Dagger Peak PM). USGS Astrogeology.*

east-west direction equal to s/√2.

The north side of the octagon is located at 1.0803°° S where the numerical value is almost exactly equal to 1/100 of the number of degrees of the internal angle of a pentagon (108). The centre of the octagon is at 5.2358°°° N (Arsia Mons Prime Latitude). The numerical value of this latitude is almost exactly equal to $2\varphi^2 = 5.2361$. The longitude of the octagon centre is 76.2500° W or 61.0000°°° W (Dagger Peak PM), and 229.2500° W (Sharonov Triangle PM).

Approximately 5 degrees north of this first octagon-shaped crater is a second octagon-shaped crater at 70.9117° E 3.9528° N (Fig 5.2). It is very well fit by an octagon which has a side dimension of 0.3000 latitude degrees and has exactly the same orientation as the first crater. Its east and west sides nicely align to linear segments of the crater perimeter. The centre of the second octagon has a latitude of 3.1623°°° N which is equal to √2√5 sacred degrees. Hence, once again, we see the use of √2 for a latitude measurement. The latitude of the south side is 10.3914°° N (Arsia Mons PL) which has a numerical value very close to $6\sqrt{3} = 10.3923$. The longitude of the octagon centre is 61.0086°°° W (Elysium Mons PM), approximately the same value as for the centre of the other octagon crater but from a different prime meridian. The western ends of the north and south sides have a longitude of 68.7496°° W (Dagger Peak PM) and 204.7496°° W (Sharonov Triangle PM). These numbers round to 68.75 and 204.75. The west side has a longitude of 68.9988°° W (Dagger Midline

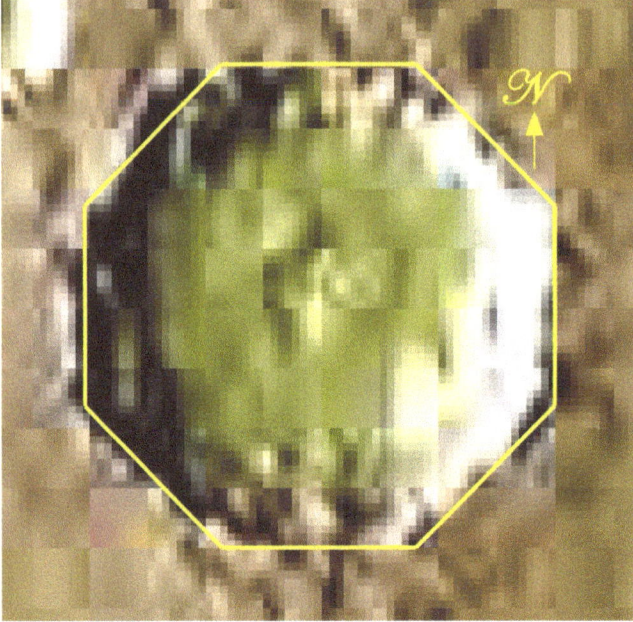

Fig. 5.2: *The second octagon-shaped crater is at 70.9117° E 3.9528° N. The side length of the fitted octagon is 0.3000 latitude degrees. The octagon aligns with the north and south sides of the crater and runs along linear segments of the east and west sides. The octagon centre is located at √2√5°°° N and 61.01°°° W (Elysium Mons PM). The west side is at 69.00°° W (Dagger Midline PM). USGS Astrogeology.*

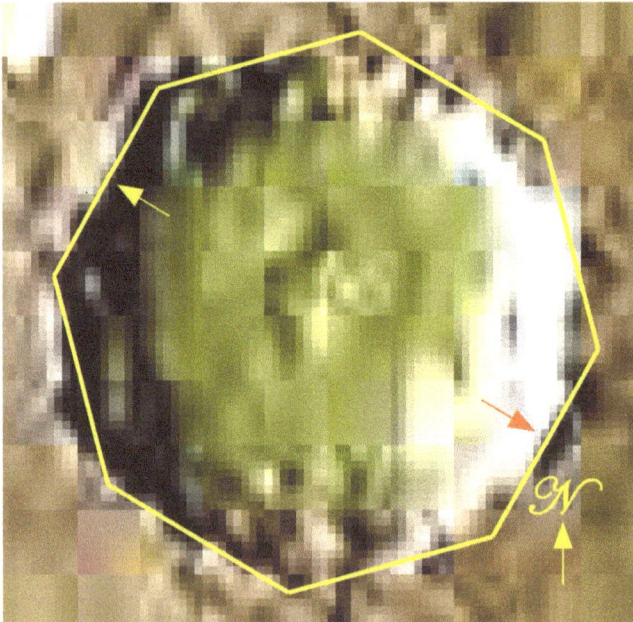

Fig. 5.3: *The octagon in Fig. 5.2 rotated counter-clockwise by 15 degrees Arrows show alignments to linear sections of the crater perimeter which were not fit by the previous octagon. The latitude of the west vertex is 4.0040° N and the latitude of the east vertex is 12.0013° N (AMPL). USGS Astrogeology.*

PM), and 204.9988°° W and 184.4990°°° W (Sharonov Tower PM), and 156.9988°° W (Pavonis Caldera PM). These numbers round to 69, 205, 184.5 and 157.

If we rotate the fitted octagon by 15 degrees in the counterclockwise direction about the same centre, we obtain the octagon shown in Fig. 5.3.

Here we see an excellent fit to 2 linear sections of the crater perimeter on the southeast and northwest sides (red and yellow arrows) and to a long section of the crater perimeter on the southwest side. The latitude of the west vertex is 4.0040° N. This value would be approximately 8° N in a 720 degree system which would reflect the number of sides in an octagon. The latitude of the east vertex is 12.0013° north of the Arsia Mons Prime Latitude (AMPL). The southwest vertex is 10.5012°° N (AMPL). The longitude of the northwest vertex is 76.5000° W and 68.0000°° W of the Elysium Mons PM, 77.5000° W and 62.0000°°° W of the Dagger Midline PM, 176.5000° W of the Pavonis Caldera PM, and 230.5000° W of the Sharonov Tower PM. The northeast vertex is at 61.5050°°° W (Dagger Peak PM) and the south vertex is at 184.2496°°° W (Sharonov Tower PM). These numbers round to 61.5 and 184.25. Finally the southwest vertex is at 77.5046° W and 62.0037°°° W (Dagger Peak PM), 176.5046° W (Pavonis Mons PM), and 230.5046° W (Sharonov Triangle PM). These numbers round to 77.5, 62, 176.5 and 230.5.

These were the only 2 craters of this type that I could locate so it is not possible to generalize about them. However, if side length was the intended unit of measurement, then an octagonal crater with a side length of 1 latitude degree could be considered to be the perfect unison note of the fundamental octave in a chromatic scale. With this assumption, the side length of the first octagonal crater would be a major 2nd note 2 octaves below this fundamental octave (9/8 divided by 2^2 = 0.2813). Likewise, the side length of the 2nd octagonal crater would be a minor 3rd in this same octave (6/5 divided by 2^2 = 0.3000). Thus the sizing of these craters seems to follow the chromatic scale in the same way as for the square, pentagon and hexagon craters.

It is interesting to note that the sum of the internal angles of a regular octagon is 1080° which is 10 times the value of the 108° internal angle of a pentagon or the angle between the star points in a pentagram. This is all the more remarkable with the finding that the north side of the first octagon is located at 1.08°° S in which the numerical value is 1/100th the size of 108. This would suggest that the octagon fits right in with major themes in the Martian architecture and is connected to the golden ratio as well as to √2.

Summary of Polygonal-Shaped Craters

The octagonal-shaped crater is the last regular polygonal shape of crater that I was able to discern. The square, pentagonal, hexagonal and octagonal craters form a unique database that can be studied in order to ascertain a part of the mindset of the ancient Martian civilization which

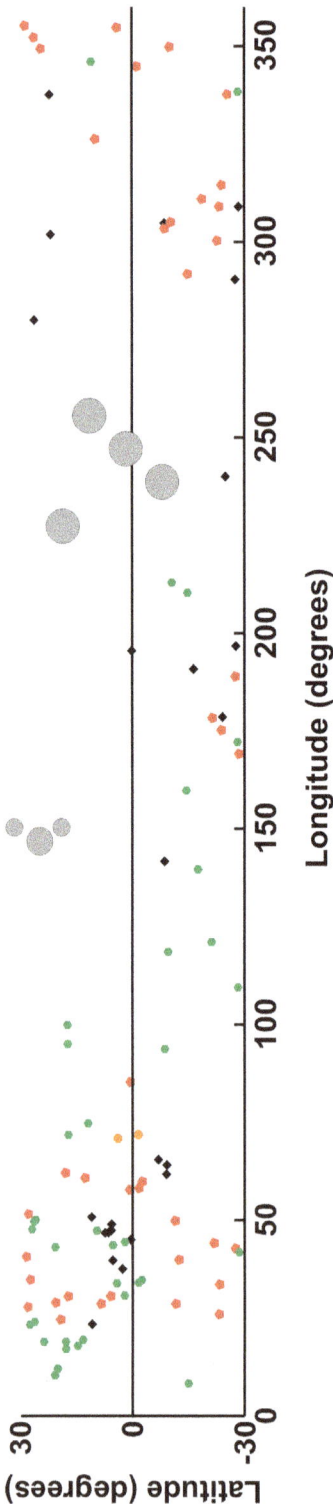

Fig. 5.4: *Distribution of polygon-shaped craters around the planet between 30° S and 30° N. "Square" craters are shown as black diamonds, "pentagon" craters as red pentagons, "hexagon" craters as green hexagons and "octagon" craters as orange octagons. The Elysium group of mountains, Olympus Mons and the Tharsis Montes are shown as grey circles.*

created them. These craters seem to share key properties which can be more clearly examined when the data are pooled. To start with, they seem to be located most densely in the region of the planet which is opposite to the major mountains of Mars. This is very well illustrated in Fig. 5.4 which shows a density map of the polygon craters. The densest cluster is in the northern hemisphere between 10° E and 75° E. Very few polygon-shaped craters are in the regions of the mountains, and the ones that do occur between the longitudes of the Elysium group of mountains and the Tharsis giants are located in the southern hemisphere except for the Nicholson Crater. Hence there seems to be a very clear intent not to interfere with the sacred geometry of the mountains, and to create a message independent of the mountains. Overall, 57 (53.8%) of the polygon craters are located north of the equator and 49 are located south of the equator.

Part of the message of the polygon-shaped craters appears to be connected to the cardinal directions of the compass, in particular, to East and to North. Most of the square-shaped craters are oriented so that their vertices point to the cardinal directions. Most of the pentagon-shaped craters point to the east and a substantial number point to the north. Most of the hexagon-shaped craters point north-south and a substantial number point east-west. The purpose for the emphasis on the east can be assumed to be tied to a reverence for the sun and perhaps stars, constellations or planets which appeared at dawn. The purpose for an emphasis on the north is more difficult to deduce, but there does seem to be a bias towards the northern hemisphere. The major mountains and their sacred geometry patterns are confined to mostly the northern hemisphere. Perhaps there was a particular northern circumpolar constellation or star which the Martians revered that was visible all year round in the northern hemisphere such as the Polaris star or the constellation Cassiopeia seen from planet earth today.

The sizing of craters according to chromatic scale intervals suggests that music played an important role in their spirituality. All 12 notes of the chromatic scale were found over a range of up to 5 octaves. Whether or not the craters actually formed a score of music is open for speculation.

The fact that elements of polygonal shapes such as vertices and centre (and the cardinal direction midpoint for the pentagonal shapes) were located at very meaningful coordinates suggests that part of their role was to create sacred geometry. A large part of that sacred geometry was centred on the pentagram and therefore the golden ratio φ. But there was also reference to the equilateral triangle and to the square as sacred geometric shapes. More mysterious is the use of several integers and integers and one-half which do not appear to be connected to sacred geometry other than being whole numbers in one degree system or another. Perhaps they fulfilled a numerological function or represented high-numbered harmonics. Nevertheless, the appearance of such numbers and small integers in coordinate values, as well as irrational numbers and numbers reflecting sacred geometric shapes, provides excellent validation for the existence of the 8 prime meridians, 2 prime latitudes and 3 systems of degrees presented in Chapter 1. These numbers would never show up with NASA coordinates based on the Airy-0 Prime Meridian except for some latitude values referenced to the equator, and the later would be restricted to the standard 360 degree system.

Participatory Sacred Geometry

With the study of the polygonal-shaped craters, we are now in a position to formulate a new concept found in Martian sacred geometry, the concept of *participatory* sacred geometry. The polygonal-shaped craters only provide part of the information required to create the geometric shape which fits them. The observer has to fill in the blanks in order to arrive at the final product. We have seen this previously in *Intelligent*

Mars I where mountain positions had to be determined from survey craters before sacred geometry patterns could be accurately drawn. Also only the Pentagram Pyramid star points and the surveyed positions of Olympus Mons and Pavonis Mons are provided for the observer to construct the circle, equilateral triangle and pentagram fitting the Vitruvian Martian. Hence, the observer has to be a co-creator. This is somewhat akin to quantum mechanics where the observer has an important impact on the state of an uncollapsed wave function. The purpose of participatory sacred geometry would seem to be to conceal the artificiality from unsophisticated observers perhaps to avoid detection from enemies. However, it may also be a fundamental aspect of the Martian spirituality which requires a participatory rather than passive role of the individual. Further evidence of participatory sacred geometry will be revealed as we go on to examine some very unique special purpose craters. But first, I want to discuss in the next chapter how craters are distributed in size across the Martian landscape.

6

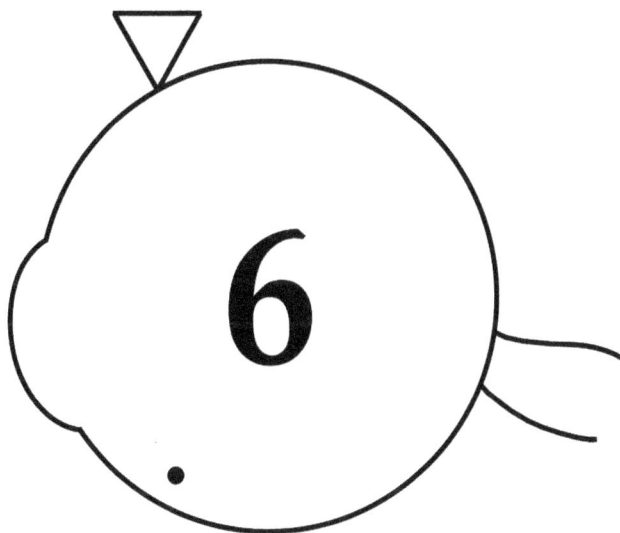

Martian Craters Come in Standard Sizes

I became interested in studying the distribution of crater sizes after I had compiled a large database of craters for the purpose of determining sacred geometry and other relationships based on their location. Not only was I interested in determining distances and bearing angles between sites but also I wanted to see if craters tended to radiate outwards in groups with a shared meaningful bearing angle from important sites such as Olympus Mons. As well, I was trying to determine whether crater placements formed some kind of grid for locating planetary coordinates from an overhead spacecraft. Since size of crater might be an important grouping factor for a given characteristic, I measured each crater's diameter at the same time that I determined its coordinates. In order to do this correctly, I had to take into account the crater's latitude since the map projections of the various USGS maps I was using distorted the size of an area in relationship to its latitude. I measured all the craters with a diameter of 5 km or greater for the region between 145° E to 330° E and between 10° S and 90° N. This region was chosen so as to centre on Olympus Mons and the 3 Tharsis Montes. I also measured a portion of the craters smaller than 5 km in diameter, and several craters outside of this region.

Since I now had a sizeable database of crater sizes, I felt that this data could be used to examine the frequency distribution of various crater sizes to see if any artificiality would be revealed. If a large proportion of the craters were the result of intelligent engineering rather than random meteorite strikes, the creators may have used a technology that favoured one or more discrete sizes rather than a smooth distribution of sizes. However, my database was not complete since I had focused on the larger craters and ignored many of the smaller ones. I decided that I would have to measure the remaining smaller sized craters in order to avoid distortions caused by a sampling bias. As it would have been too great a task to do this for the entire region mentioned above, I limited myself to all craters north of the equator between 225° E and 270° E. This region covers about 9 million square km and hosts the largest mountains except Arsia Mons, the western 40% of Olympus Mons and a small portion of Pavonis Mons south of the equator. I found 1590 craters in this region ranging in size from 2 km to 62 km in diameter.

Fig. 6.1 shows a plot of the number of craters of a given size found in the 9 million square km region north of the equator between 225° E and 270° E. Note that a logarithmic scale is used for both the x and y axes. Each data point represents the number of craters found in a 0.67 km interval of crater sizes. Thus the highest point on the graph represents all of the craters found in the region which have a diameter ranging from 3.33 km to 4.00 km in size. Conventional science tells us that the frequency of occurrence of craters on Mars is related to crater size in an inverse log-log relationship. In simple language, this means that the larger-sized craters are less numerous than the smaller craters such that if you plot the number of craters of a given size (y-axis) against crater diameter (x-axis) and make both the x and y axes logarithmic scales, you will get an approximately straight line with a negative slope. Fig. 6.1 shows conformity to the straight line relationship for craters from 5.33 km in diameter to 26.67 km in diameter with a correlation coefficient of 0.958. The slope of the line is -3.059 log number of craters per log crater diameter in km. For craters smaller than 3.33 km in diameter, the straight line relationship breaks down dramatically and crater counts go down instead of up. The reduced crater counts for the smaller-sized craters would be explained by conventional science as resulting from the craters being erased over time due to resurfacing events or continual erosion. The final 6 data points use a larger interval size than 0.67 km so as to smooth out the data. To maintain the same scale as the rest of the graph, the number of craters in these larger intervals is then divided by the factor by which the interval was increased (e.g., for an interval size of 1.33 km, the number of craters is divided by 2). There are also 11 craters larger in diameter than shown on the graph, but they are not numerous enough to produce a number count equivalent to at least 1 crater per interval of 0.67 km in crater diameter size.

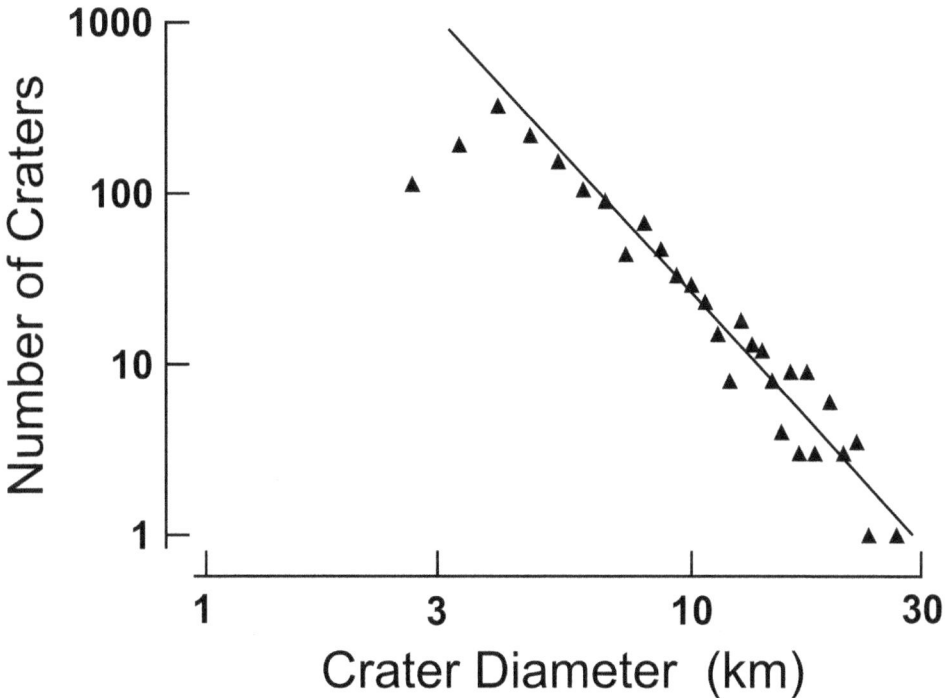

Fig. 6.1: *Plot of crater frequency (# of craters in a 0.67 km interval of crater sizes) against crater size for 1579 craters between 225° E - 270° E and 0° - 90° N. Both axes are on a logarithmic scale. The straight line is a fit to all the data points for craters having diameters ranging from 5.33 to 26.67 km. The slope of the curve is -3.059 log number of craters per log crater diameter (km), and the correlation coefficient of the fit is 0.958.*

Since the logarithmic scale on the y-axis compresses the data at higher values, information tends to get lost. I therefore re-plotted the data from Fig. 6.1 using a linear scale for both the x and y axes (Fig. 6.2). By doing so, a very large and fairly symmetrical peak appears at the 3.33 - 4.00 km data point which dominates all the other data. The truth of the matter is that most of the craters 2 km or larger in diameter in this region of the planet fall within a fairly narrow size range, with over 60% occurring in the first peak alone. Note also that there is now a second smaller peak that shows up at the 7.33 – 8.00 km data point which is approximately double the diameter size of craters seen in the large peak. This second peak appears to be simply noise in Fig. 6.1, so it is completely lost when a log scale is used. Since the peak at 7.33 - 8.00 km is relatively small, it might be explained away as being simply due to a random burst of meteorites of a set size hitting the planet. But the fact that the diameter sizes for the 7.33 - 8.00 km peak are close to twice the diameter sizes for the 3.33 - 4.00 km peak makes one wonder if some sort of intelligent

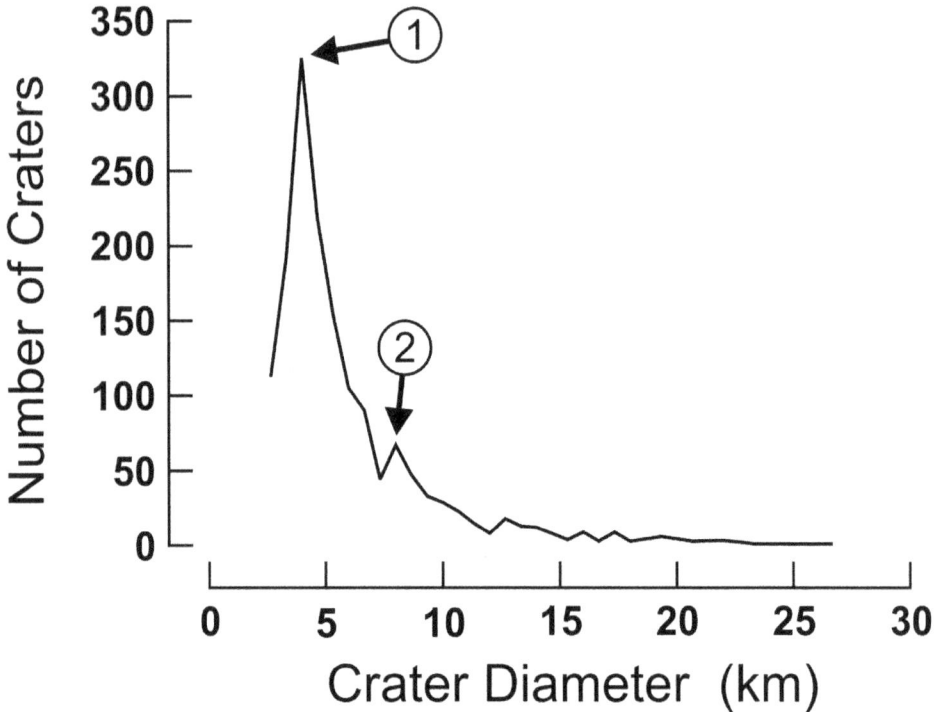

Fig. 6.2: *The data in Fig. 6.1 re-plotted using a linear scale for the x and y-axes instead of a logarithmic scale. A very large symmetrical peak (1) occurs at crater diameters of 3.33 - 4.0 km. A smaller peak (2) occurs at crater diameters of 7.33 - 8.0 km.*

process was behind the occurrence of higher numbers of craters in the second peak. This can be tested by examining the distribution of crater sizes from a larger region. If the second peak was due to randomness, data from a larger region should even out the curve and make this peak smaller in relationship to the rest of the curve.

Having demonstrated the existence of 2 peaks with an unbiased sample of craters from the smaller region of the planet, we can now look at my entire database of craters which includes a much larger area where I did not measure all of the craters less than 5 km in diameter. This expands the total number of craters from 1590 to 4641 and covers more than 1 quarter of the planetary surface. A plot of these data is shown in Fig. 6.3 in which the number of craters in 0.67 km wide intervals of diameter sizes (y-axis with a log scale) is plotted against crater diameter (x-axis with a log scale) similar to the plot in Fig. 6.1. Fig. 6.3 also shows conformity to the straight line relationship but now the range of crater sizes that fit the curve extends from 9.33 km in diameter to 65.33 km in diameter. The slope of the line is -2.594 log number of craters per log

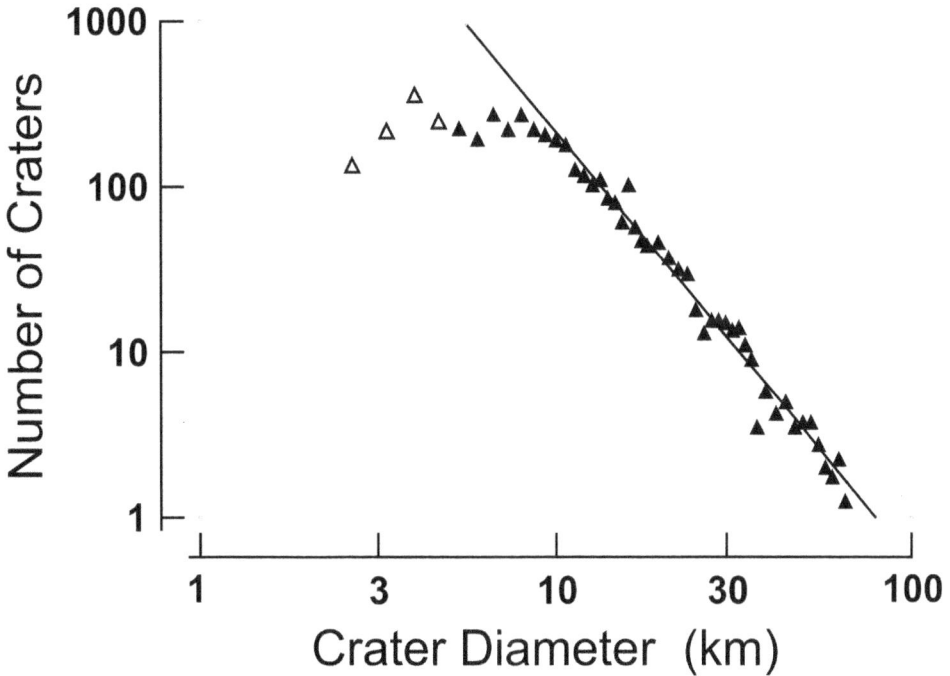

Fig. 6.3: *Plot of crater frequency (# of craters in a 0.67 km interval of crater sizes) against crater size for 4634 craters between 145° E - 330° E and 10° S - 90° N. Both axes are on a logarithmic scale. The straight line is a fit to all the data points for craters having diameters ranging from 9.33 km to 65.33 km. The slope of the curve is -2.594 log number of craters per log crater diameter (km), and the correlation coefficient of the fit is 0.989. Unfilled triangles represent incomplete data.*

crater diameter in km and the correlation coefficient is very strong at 0.989. The deviation from a straight line relationship now breaks down markedly for crater sizes smaller than 9.33 km rather than for crater sizes smaller than 3.33 km as was seen in Fig. 6.1.

When the data in Fig. 6.3 are re-plotted using a linear scale for both the x-axis and y-axis, (Fig. 6.4), the 2 peaks in Fig. 6.2 not only are larger but also are accompanied by 3 additional peaks. Although the data for the first peak are now incomplete since not all of the craters less than 5 km in diameter were measured in the expanded region, the peak is so large that its presence is established beyond question. Only its true magnitude is not fully established for the larger region of the planet. The size of the peak at 7.33 – 8.00 km (peak #2 in Fig. 6.2 which is now peak #3 in Fig. 6.4) is much more pronounced in Fig. 6.4 thus confirming that it was not due to a random deviation from the overall shape of the rest of the curve in Fig. 6.2. In addition, there is now a sizeable peak at 15.33 - 16.00 km

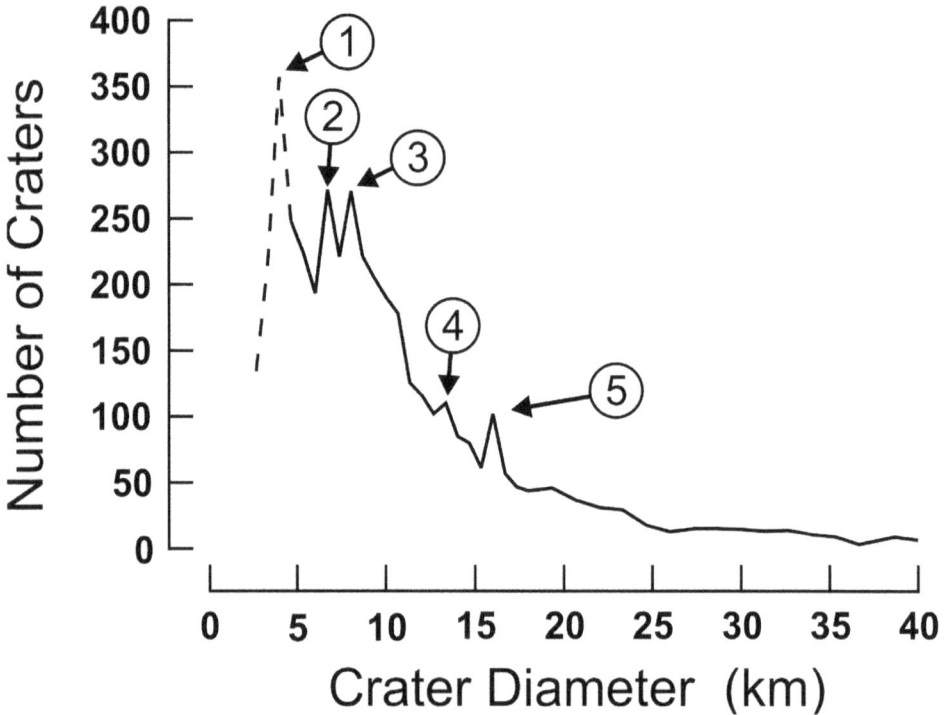

Fig. 6.4: *The data in Fig. 6.3 re-plotted using a linear scale for the x and y axes instead of a logarithmic scale. Sizeable peaks occur at crater diameters 3.33 - 4.00 km (1), 6.00 - 6.67 km (2), 7.33 - 8.00 km (3), 12.67 - 13.33 km (4) and 15.33 - 16.00 km (5). The curve has been truncated at a crater diameter size of 40 km to better display the peaks of interest. Dashed portion of curve represents incomplete data.*

(peak #5) which is approximately twice the crater diameter size for peak #3, and 4 times the crater diameter size for peak #1. This provides extremely strong evidence that a substantial number of the craters have been created with standard sizes on Mars by an advanced civilization rather than all craters being created by asteroids or comets hitting the planet over billions of years. This conclusion is further reinforced by the crater diameters found at peaks #2 and #4 in Fig. 6.4. The second peak occurs at crater diameters ranging from 6.00 km to 6.67 km and the 4th peak occurs at crater diameters which are approximately twice this size, ranging from 12.67 km to 13.33 km.

Before concluding this chapter I would like to mention that I excluded a type of crater from my crater databases which is known as a pit crater. The upper picture of Fig. 6.5 shows the Ceraunius Catena, a chain of craters located on Alba Mons at about 251.9°E 37.1°N. The crater pointed out by the lower arrow has an oval shape and the crater pointed out by

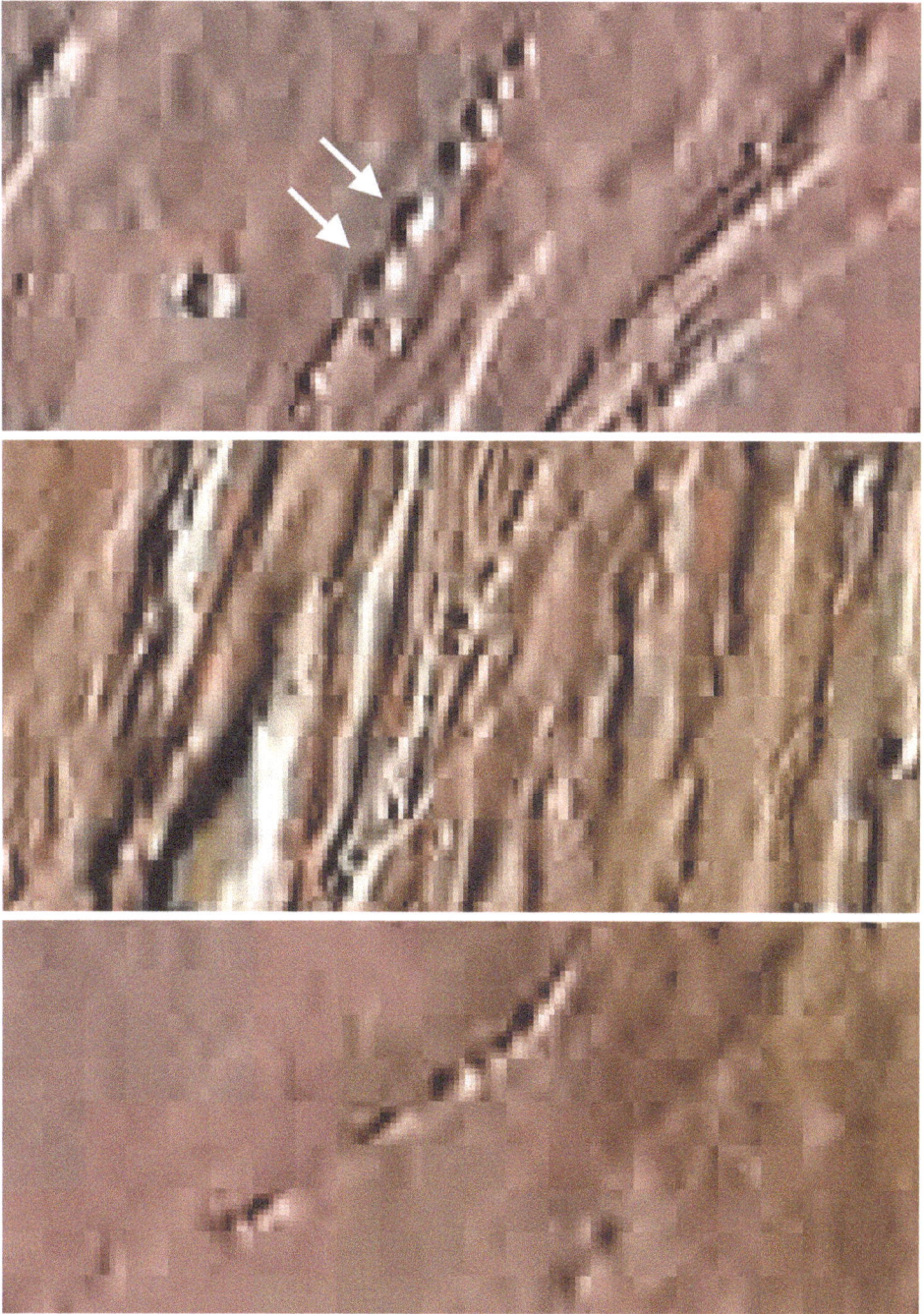

Fig. 6.5: *Examples of pit craters on Alba Mons. These craters are believed to have arisen from the collapse of terrain into caverns below the surface rather than from meteorite strikes. Top picture is of the Ceraunius Caterna (251.9°E 37.1°N). The middle picture (≈251.8° E 31.5° N) and bottom picture (≈254.0° E 43.7° N) show craters which mostly merge to form continuous channels. USGS Astrogeology.*

the upper arrow seems to be composed of 2 overlapping craters. The diameter of all the craters is approximately the same, about 4.7 km except for the oval shaped crater. The long axis of this crater has a diameter of about 6 km. Their size is slightly larger than the 3.33 – 4.00 km peak found in Figs. 6.2 and 6.4. I also wish to point out that the southern craters in the chain merge to form a trench. The conventional explanation for pit craters is that they are formed by surface collapse rather than by meteorite impact. One of the ways in which they are believed to arise is by the collapse of sections of a lava tube which is a tube formed by lava whose surface layer hardened while molten lava continued to flow underneath. When the supply of lava was exhausted an empty tube remained which could be quite long. This could explain both the linearity of the chain of craters and homogeneity in size of the craters since the tube would have a consistent diameter over its length. The middle and lower pictures of Fig. 6.5 are other examples of pit craters which are very close together forming continuous chains. I avoided putting craters into my database which formed chains or were part of linear trenches since these craters were likely to belong to the category of pit crater.

In summary, craters on Mars do not seem to have resulted solely from random bombardment of the planet with meteorites over the ages. There is a strong possibility that many, if not most, of the craters on Mars have been artificially created using a process that expends standard amounts of energy to produce standard sizes of craters. The purpose of the plethora of 3.33 - 4.00 km diameter craters was probably to camouflage the presence of their civilization on the planet, giving the impression that it was a barren place without much of a barrier to the bombardment of the surface by passing meteorites. It certainly has worked well on us, with very few even suspecting that Mars was home to an advanced civilization. But why did they make them that particular size? In keeping with everything else that has been found out about the sites on Mars, there must be an explanation based on sacred geometry, numerology or spirituality. After testing out a few hypotheses, I found that if the true peak within the 3.33 – 4.00 km diameter interval occurred at 3.9516 km, it would be exactly equal to 1/15 of a degree measured at the equator. This would mean that if you placed 5400 of them end-to-end around the equator, you would exactly circle the planet. Now 5400 is 100 times 54, and 54° is 1/2 the internal angle of a regular pentagon or of the angle between the star points in a pentagram. If the crater radius for this size of crater is used instead of the diameter, then the planetary circumference at the equator would equal 10800 crater radii or 100 times the number of degrees of the internal angle of a pentagon. Also the sum of the internal angles of a pentagon is 540°. Therefore, this crater diameter (or radius) could ultimately be a powerful reference to the golden ratio. The use of binary multiples of this size found for peaks #3 and #5 in Fig. 6.4 would maintain this relationship by having 3.9516 km as a basic factor. Similarly for peak #2 in Fig. 6.4, if the true peak within the 6.00 – 6.67 km diameter interval occurred at 6.5941 km, the number of craters of this size which could fit around the equator would be 2000φ = 3236 craters, and a crater diameter twice this size would fit 1000φ craters around the equator. Higher resolution measurements are required to permit a test of these hypotheses.

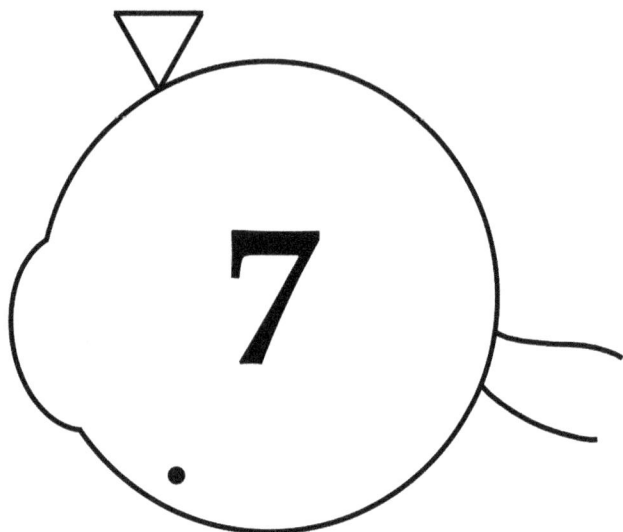

The Denning and Savich Craters

In *Intelligent Mars I*, we have seen how craters have been used as survey markers to triangulate the original positions of the Martian mountains. Some have also been shown to be at sacred distances and to form sacred bearing angles with several of the mountains and other important sites. The previous chapters in this book have demonstrated that many craters are shaped in the form of regular polygons that mark out important longitudes and latitudes, and are sized according to notes in the chromatic scale. There are also groups of craters which have standard diameter sizes that are in binary relationship to each other, e.g. approximately 4, 8 and 16 km. In next few chapters, I am going to show you several craters that have specialized functions, some of which are truly startling in their sophistication and purpose. I will start off with the analysis of the Denning and Savich Craters in this present chapter. To the casual observer there does not appear to be anything unusual about these craters. However, closer examination tells a very different story.

The Denning Crater

The Denning Crater (33.4934° E 17.4356° S) is a huge crater with a diameter of about 160 km. The longitude of the crater is 101.9369°° W (Dagger Midline PM) with a numerical value very close to $63\varphi = 101.9361$. Its latitude is 15.4983°° S which is close to 15.5000°° S. This crater grabbed

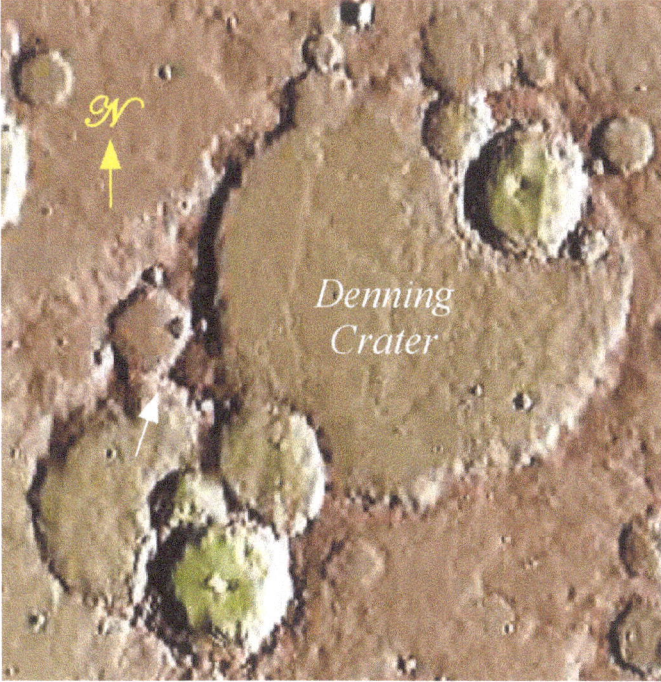

Fig. 7.1: *The Denning Crater has a large cluster of craters at its southwest corner and a second sizeable cluster of craters in its northeast region. The white arrow points to a wedge-shaped outcropping of the western perimeter of the Denning Crater. USGS Astrogeology.*

my attention due to a curious large notch shaped like an arrowhead which protrudes outwards from the west side of the crater perimeter (Fig. 7.1). The northwest side of the notch is linear and has a bearing angle of 36° in the clockwise direction (Fig. 7.2). The southwest side of the notch is also linear for the western half of its length, and this portion has a bearing angle of 60° in the counterclockwise direction. The angle at the point is 84°. What is most interesting, however, is that the point of the arrowhead

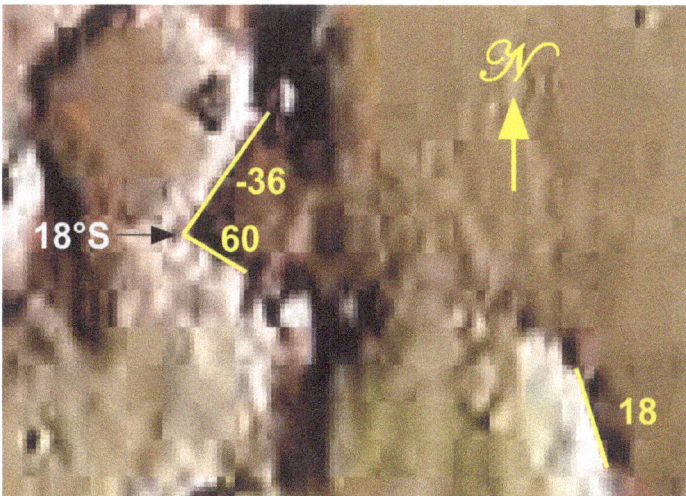

Fig. 7.2: *The point of the wedge in the western perimeter of the Denning Crater has a latitude of exactly 18° S. The bearing angle of a linear section of the perimeter of a crater to the southeast is also 18° (counterclockwise). USGS Astrogeology.*

is at exactly 18.0000° S. The northeast perimeter of a rather large crater just to the southeast of the arrowhead has a bearing angle of 18° in the counterclockwise direction, the same value as the latitude of the arrowhead point. The strong presence of the numbers 18 and 36 are suggestive of the size of the angle of a pentagram star point which is 36 degrees. It should be noted that 18.0000° S is also 16.0000°° S, and it is 9.9004° S with reference to the Arsia Mons PL. The latter has a numerical value which is extremely close to $7\sqrt{2} = 9.8995$. The value of 16 is a binary number, and both 16 (= 4^2) and $\sqrt{2}$ could refer to the geometry of a square.

The small crater which borders the upper part of the arrowhead has 2 other linear sides, one with a counterclockwise bearing angle of 48° and the other with a clockwise bearing angle of 45° (Fig. 7.3). Another arrow-shaped structure is within its interior and a 3rd arrow-shaped structure lies on the north side of its perimeter. The sides of both of these 2 arrow-shaped structures, which seem to be depressions in the landscape, have bearing angles of ±54° which further reinforces the pentagram theme since this is one-half the size of the angle between the pentagram star points or of the angles in the interior pentagon of the pentagram. Also, the angle formed by the sides of each arrow tip is 72° which is the size of the angles at the base of a pentagram star point. The coordinates of the tip of the arrowhead inside the crater are 31.8832° E 17.7027° S. The longitude is 172.1772 = $77\sqrt{5}$°°° W (Pavonis Mons PM) and the latitude is 8.5361°° S (AMPL) with a numerical value close to $e\pi = 8.5397$. The coordinates of the tip of the arrowhead just north of the small crater are 31.7270° E

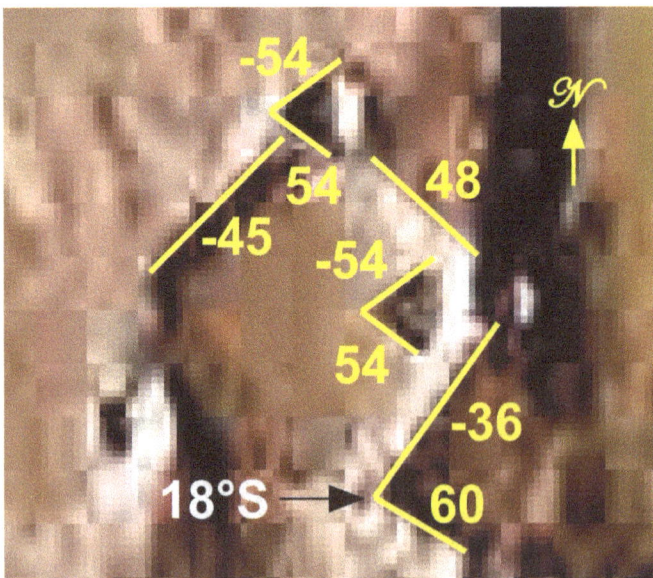

Fig. 7.3: *Two more arrowheads north of the Denning Crater arrowhead. The perimeter of the small crater north-west of the Denning Crater arrowhead has 2 additional straight line segments with bearing angles of 45° in the clockwise direction and 48° in the counter-clockwise direction. USGS Astrogeology.*

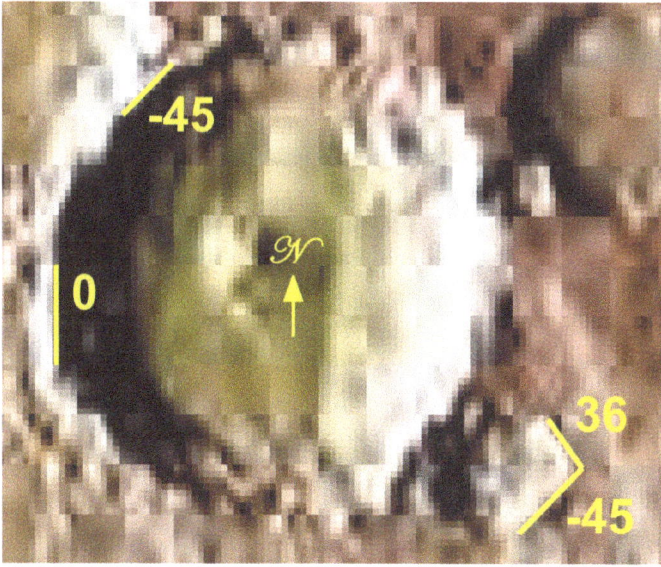

Fig. 7.4: *The west perimeter of the large crater in the northeast section of the Denning Crater has a bearing angle of 0°. A linear section of the crater perimeter on the northwest side has a clockwise bearing angle of 45°. A small crater southeast of the larger crater has linear sides which form an angle that points due east. USGS Astrogeology.*

17.3817° S. The longitude is 103.4468 = 14(e²)°° W (Dagger Peak PM) and the latitude is 15.4504 = atan(1/(√5φ))°° S whose numerical value is the same as the degrees of rotation of the pentagram pyramid from having a star point aimed due north, and is also close to the longitude of the pentagram pyramid from the Pavonis Mons PM (see Fig. 1.10).

The west perimeter of the large crater in the northeast part of the Denning Crater (Fig. 7.4) has a linear section with a bearing angle of 0° and hence points to the poles of the planet. A second linear section of the crater perimeter on the northwest side has a clockwise bearing angle of 45°. To the southeast of the large crater is a much smaller crater with linear sides which form an asymmetrical arrowhead with an angle of 99° pointing due east. The bearing angles of its sides are 36° and -45°. The coordinates of the arrowhead's point are 34.8629° E 17.1888° S. Its latitude is 8.0793°° S (Arsia Mons Prime Latitude) which has a numerical value close to 5φ = 8.0902. Its longitude is 213.0476° W (Sharonov Tower PM) which has a numerical value close to 123√3 = 213.0422.

All of the bearing and arrowhead angles connected with the Denning Crater and its associated craters are divisible by 3. The purpose of the Denning Crater arrowhead seems to be to mark the latitude of 18° S which is a reference to the pentagram. The other 3 arrowheads also reference the pentagram a) in the bearing angles of their sides, b) in 2 of the arrowhead angles, and c) in some of their coordinate values. Note that all of the arrowheads point either due west or due east. The occurrence of so many arrowheads in this location suggests that there is much yet to learn about the Denning Crater and its associated craters.

The Savich Crater

Approximately 3500 km to the southeast of the Denning Crater is the Savich Crater. It has a diameter of about 180 km, making it one of the largest craters on Mars. It is very difficult to determine the exact coordinates of the centre of the Savich Crater since its perimeter is poorly defined (Fig. 7.5). After several attempts, I came up with a fit of its perimeter to a circle whose centre I used to provide the approximate coordinates of 96.13° E 27.52° S for the Savich Crater centre. On the western perimeter of the crater there is an inward projection shaped like an arrowhead (pointed out by the white arrow in Fig. 7.5) which appears to have been formed by the perimeters of 2 smaller craters in contact with each other. The northern small crater has a central peak, and about half of its area lies inside the Savich Crater. The southern small crater lies completely within the Savich Crater and appears to be highly eroded. The arrowhead region itself forms an enclosure consisting of depressed terrain bounded by a ridge line which forms much of its western and southern perimeter. The enclosure is almost quadrilateral in shape and has linear sections on all 4 sides of its perimeter, 2 of which form the arrowhead

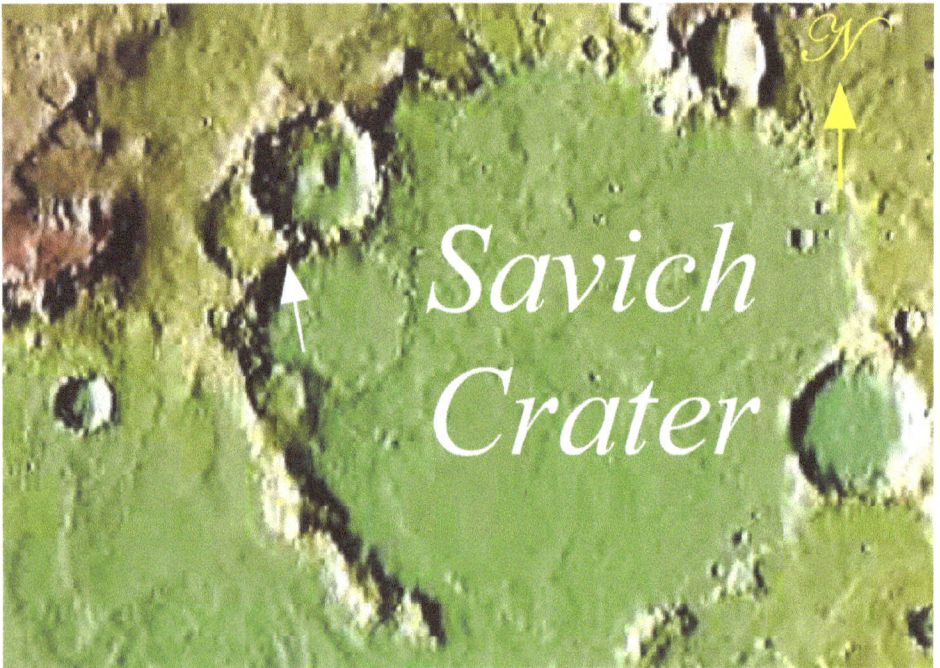

Fig. 7.5: *The Savich Crater has a diameter of about 180 km. There is a curious inward projection of its perimeter on the west side which is shaped like an arrowhead (white arrow). USGS Astrogeology.*

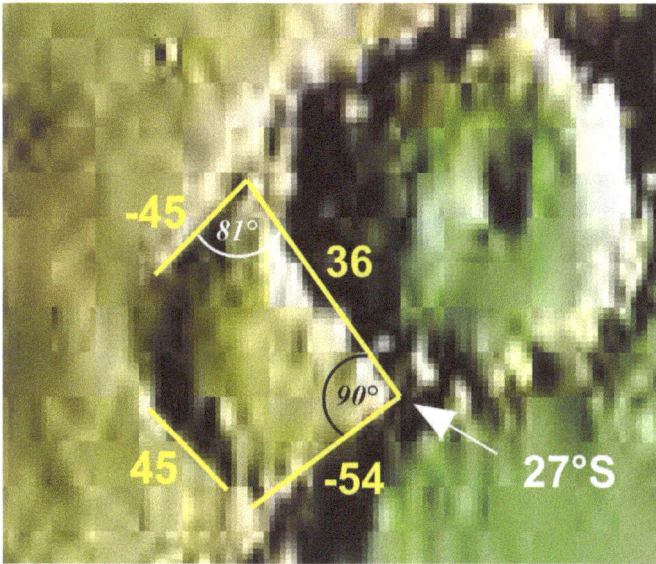

Fig. 7.6: *The arrowhead pointing eastwards into the Savich Crater creates a depressed region whose perimeter has 4 linear sections. The bearing angles of the 2 sides of the arrowhead reflect the pentagram as does the latitude of the arrowhead tip which is 27° S. The bearing angles of 45° and the arrowhead angle of 90° probably reflect the square or rectangle. USGS Astrogeology.*

pointing eastwards (Fig. 7.6). The northern side of the arrowhead has an overall bearing angle of 36° in the counterclockwise direction, and the southern side of the arrowhead has a bearing angle of 54° in the clockwise direction. The 2 sides of the arrowhead meet at a 90° angle and the meeting point occurs at the latitude of exactly 27.0000° S. The numbers 27, 36, and 54 are all associated with the pentagram since 36 is the number of degrees in a star point and 108 is the number of degrees between the star points and for the internal angles of the enclosed pentagon. The number 54 is one-half of 108 and 27 is one-quarter of 108. The northwest and southwest sides of the arrowhead enclosure have bearing angles of 45° (clockwise and counterclockwise respectively). The former meets with the northeast side of the arrowhead at an angle of 81°. Since 27 is the 3rd power of 3 and 81 is the 4th power of 3, we have here a powerful representation of the number 3 in this seemingly insignificant but truly remarkable enclosure. The 3rd power might represent the equilateral triangle and the 4th power, the square. It is likely that the bearing angles of 45° and the arrowhead angle of 90° reflect the square or rectangle.

If the southern edge of the arrowhead in Fig. 7.6 is extended to the southwest, it passes directly through the centre of a smaller crater to the southwest (Fig. 7.7). The perimeter of the smaller crater has straight line sections on its northern and southern sides which, if extended to the northeast, pass through the tip of the arrowhead. The northern line has a clockwise bearing angle of 60° and the southern line has a clockwise bearing angle of 48°. Together, they create an angle of 12° at the arrowhead tip. The coordinates of the smaller crater are 93.3840° E

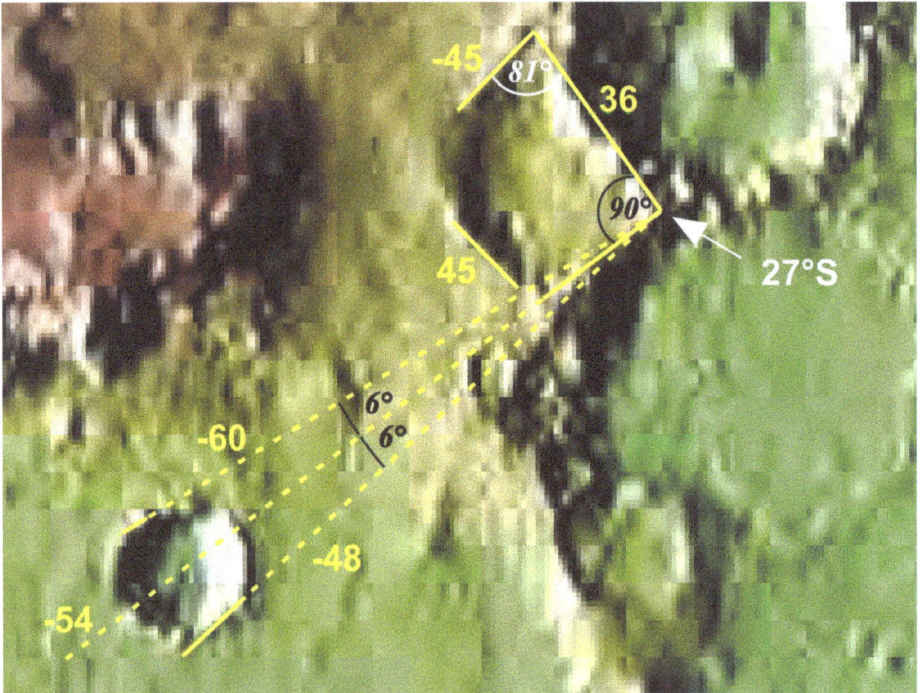

Fig. 7.7: *When the lower edge of the arrowhead in Fig. 7.6 is extended to the southwest, it passes directly through the centre of a small crater. Two straight line sections of the small crater's perimeter focus on the arrowhead tip to create an angle of 12°. USGS Astrogeology.*

27.7748° S. The latitude is 22.2198°°° S which has a numerical value close to atan(1/(√2√3)) = 22.2077. The longitude is 153.7207° W or 122.9766°°° W (Pavonis Mons PM). These numbers are close to 95φ = 153.7132, and 71√3 = 122.9756 or 76φ = 122.9706.

In the region immediately west of the arrowhead enclosure there is another area of depressed terrain whose northwest and southwest linear sides have bearing angles of 45° in the clockwise and counterclockwise directions respectively (Fig. 7.8 left). The southeast perimeter of a small crater (Fig. 7.8 top left) is continuous with the northwest side of the depressed terrain. Just south of this crater is another very shallow crater whose northeast side has a linear section with a bearing angle of 45° in the counterclockwise direction. When the line fitting the northeast side of this crater is extended southeast, it aligns perfectly with the southwest linear side of the arrowhead region. Finally, the northwest perimeter of the shallow crater inside the Savich Crater (Fig. 7.8 bottom right) has a linear section with a clockwise bearing angle of 45°. When extended to the northeast, this line intersects the tip of the Savich Crater arrowhead,

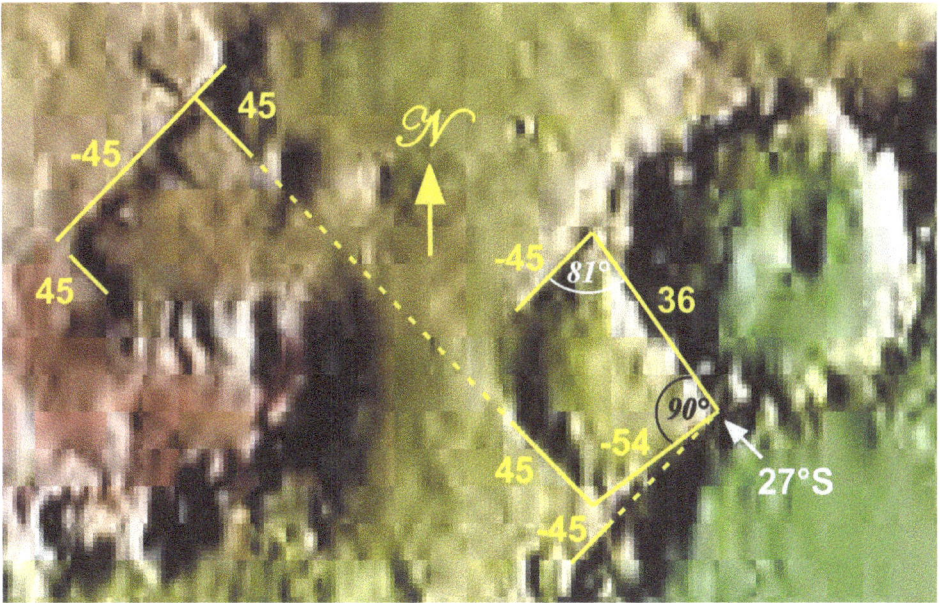

Fig. 7.8: *A region of depressed terrain west of the arrowhead is bounded by edges having a clockwise or counterclockwise bearing angle of 45°. USGS Astrogeology.*

forming an angle of 9° with the bottom edge of the arrowhead. This angle value is 1/4 the size of the angle of a pentagram star point. The high degree of emphasis on 45° for the bearing angles of straight lines associated with this region of depressed terrain suggests that it is honouring the square or rectangle since 45° is 1/2 the size of their 90° angles.

There is another way in which the Savich Crater marks an important latitude. To see this, we need to have a closer look at the small crater with the central peak whose southwest side forms the northern side of the arrowhead. There are a number of other straight line segments forming this crater's perimeter (Fig. 7.9). On the east side, moving from south to north there are straight line segments with bearing angles of 18° in the clockwise direction, and 7.5° and 30° degrees in the counterclockwise direction. The number 18 fits with the pentagram theme because it is one-half the size of the angle of a pentagram star point. Now let's look at the linear segment of the perimeter forming the crater's northwest side. It has a bearing angle of 60° in the clockwise direction, so it is matched with the 30° found for the bearing angle of the northeast side linear segment of the crater perimeter, and a right angle is formed when these 2 segments are extended to a common meeting point. It gets more interesting if the northwest line segment is extended even further so that it contacts the northern end of a dark region jutting up northwards from the Savich

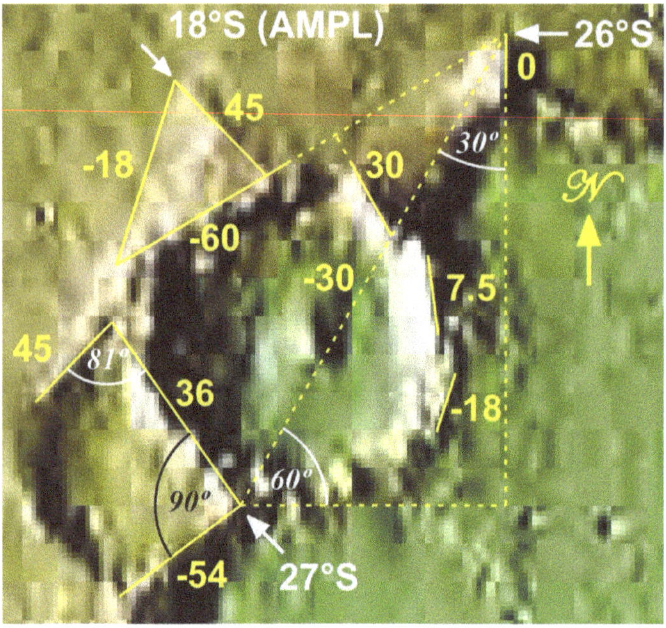

Fig. 7.9: *The perimeter of the crater on the north-west edge of the Savich Crater has many linear sections. Its extended northwest linear section intersects at 26° S with the dashed meridian line aligned to the western edge of a dark brown northwards extension of the Savich Crater peri-meter. The northern tip of a triangular-shaped region on the northwest side of the smaller crater is located at 18° S (Arsia Mons Prime Latitude). USGS Astrogeology.*

Crater. If a meridian line is drawn at the western edge of this dark region, it intersects the line which we just extended out from the northwest perimeter of the small crater (with the central peak). It does so at the latitude of 26.0000° S. The meridian line marks an important longitude as well, namely 47.0000°° W of the Dagger Peak PM. So here we have markers for both 26° S and 27° S associated with the Savich Crater. Now if you join these 2 latitude marker sites with a line, the line is found to have a bearing angle of 30° in the clockwise direction. By drawing a latitude line at the marker for 27° S so that it intersects with the meridian line at 47°° W of the Dagger Peak PM, a standard 30 - 60 degree triangle is formed with the height equal to 1 degree of latitude (59.27 km). The base length is 34.08 km which is very close to R/100 = 33.96 km. The hypotenuse is 68.45 km which is close to R/50 = 67.92 km. [Note that due to the curvature of the planet, the length of the hypotenuse is not exactly twice that of the base as would be the case on a flat 2 dimensional surface.] This triangular arrangement would enable an overhead spacecraft to obtain a direct measurement of 1 degree of latitude as well as measure its altitude and obtain longitudinal and latitudinal bearings without the need for bouncing signals off the planet or receiving positional signals from a satellite or ground-based transmitter. The advantage of this is that the spacecraft could maintain electromagnetic silence and not reveal its presence to a potential threat from outer space.

Yet another marker of an important latitude occurs in association with the northwest side of the small crater with a central peak. It is an elevated landmass in the shape of a triangle (Fig. 7.9). The northeast and northwest sides of the triangle meet at a location which is 18.0000° S and 16.0000°° S of the Arsia Mons Prime Latitude. This gives a high degree of credibility to the use of the survey centre of Arsia Mons for a prime latitude marker in addition to the equator. Note the use of the number 18 again. It would be 36 in a 720 degree system. The longitude of the northern vertex of this triangle is 52.7490° W (Elysium Mons PM), 53.7490° W and 42.9992°°° W (Dagger Midline PM), and 206.7490° W (Sharonov Tower PM). When rounded to 2 decimal places these numbers become 52.75, 53.75, 43.00 and 206.75. Most interesting is that the ratio of the length of the triangle side with a bearing angle of 45° to that of the side with a bearing angle of -18° is extremely close to the natural log of 2 (0.6926 vs. 0.6931).

In summary, all of the bearing angles (except 7.5°) and the angles at the intersection points of lines aligned to straight line segments of craters and other landforms in the vicinity of the Savich Crater near or at its western end are divisible by 3. There are several themes in addition to the number 3 theme such as the pentagram theme, the 30-60 degree theme and the 45 degree theme. The 30-60 degree triangle is 1/2 of an equilateral triangle and is associated with √3 since the height of an equilateral triangle is √3 times 1/2 of its side length. The angle of 45 degrees is associated with the square since the diagonal of a square creates 45 degree angles. It is also associated with √2 since sin(45) = cos(45) = 1/√2, and the diagonal length of a square is √2 times its side length. Several interesting coordinate markers were found which point out the locations of very meaningful latitudes and also longitudes of lessor importance which are either sacred geometry formulae or whole integers in at least one of the degree systems likely to have been used on Mars. The results from this analysis of the Savich Crater serve to point out the use of extended lines to intersect at an important location rather than marking the location with an obvious structure. This helps to camouflage the coordinate marking function of architectural constructs. It also camouflages the marking out of the distance covered by 1 degree of latitude. Thus the Savich Crater is a splendid example of participatory sacred geometry which requires the observer to complete the picture from partial data which give the necessary starting information.

The Denning and Savich Craters both refer strongly to the pentagram. With their respective arrowheads, the Denning Crater emphasizes the angles of the star points (18° being 1/2 of 36°) whereas the Savich Crater emphasizes the angles between the star points or the angles of the interior pentagon (27° being 1/4 of 108°). In addition, the Savich Crater refers to 18° S (AMPL) with the northern tip of a triangular structure to the northwest.

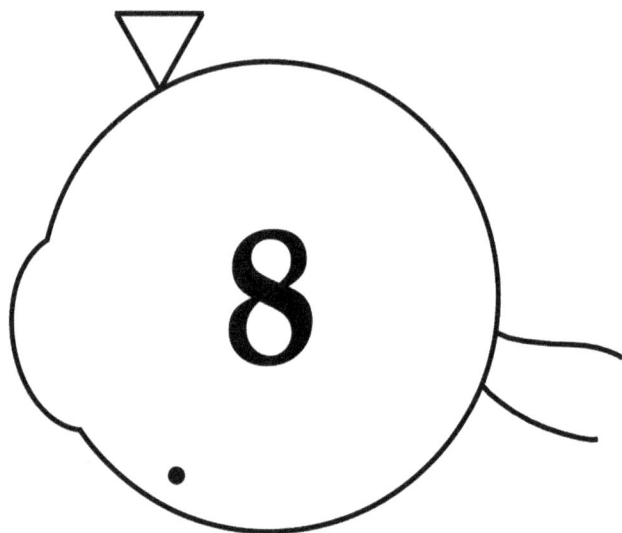

Janssen's Wheel

The next crater to be examined is the Janssen Crater which is located at 37.6152° E 2.7096° N. It is one of the largest "square" craters that I measured out (Fig. 8.1) having a diameter of about 150 km. It is named after the French astronomer Jules Janssen who, along with Joseph Lockyer, is known for the discovery of helium. I could fit a circle quite well to this crater and the coordinates that I just gave are those of the centre of the fitted circle. However, there were several linear segments of the crater perimeter which led me to suspect that a square might fit it as well. I decided to give this a try.

As with all large "square" craters, one has to experiment first in order to determine the square which best fits the perimeter of the crater. The centre, the size and the angle of rotation all have to be selected from several possibilities, each of which has its own merits. Since 45° is the most common rotation for "square" craters, I started there. I then noticed that the latitude of the crater centre was very close to $e = 2.7183°$ N so I set the latitude of the centre of the square to this value. That left only the longitude and the size of the square to determine. After a good deal of testing, I decided to settle on the square shown in Fig. 8.2. The diagonal of this square was set to 3.75 latitude degrees and the longitude of the centre (37.5495° E) was adjusted so that the western vertex lay at exactly 100°° W of the Dagger Midline PM. This is the square that I used in Table 2.1 to represent the Janssen crater since it fits a long linear segment of the crater

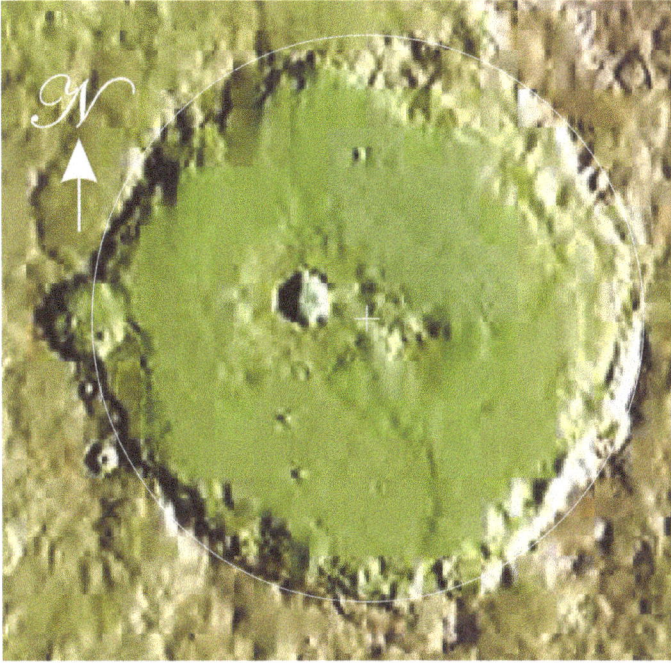

Fig. 8.1: *Based on the centre of a circle fit to its perimeter, the Janssen Crater is located at 37.6152° E 2.7096° N. Although it has an overall circular shape, it also appears to be square-like, and there are straight line segments which suggest that a square might be well fit to segments of the perimeter. USGS Astrogeology.*

Fig. 8.2: *Square with diagonal size = 3.75 latitude degrees rotated 45° from sitting flat on a latitude line. Arrows mark alignments to landmarks. The red line marks the alignment to a long linear section of the crater perimeter on the southeast side. Inset at bottom right shows alignment to 3 small craters near the southern vertex of the square. A white cross marks the square's centre located at 37.5495° E 2.7183° N. USGS Astrogeology.*

perimeter (red line, southeast side in Fig. 8.2). The northeast side touches the perimeter at several locations and is in alignment with the northeast edge of a small crater adjacent to the Janssen Crater (arrow, northeast side). The northwest side touches the outside edge of the perimeter apron at one point (arrow, northwest side). The southwest side runs along the southwest edge of the central peak of a small crater (arrow, southwest side left), and also runs along the edges of the dark regions of 3 small craters, 2 of which are very tiny, near the south vertex (arrow, southwest side right; inset bottom right). The crater perimeter exceeds the limits of the square for only 2 small outcroppings on the northeast side and 2 on the southeast side. The coordinates of this square are quite remarkable. As mentioned, the centre of the square is at a latitude of e = 2.7183° N. Its longitude (88.4983°°° W, Dagger Midline PM) is about 6 seconds less than 88.5000°°° W as are the longitudes of its north and south vertices. No less remarkable is the longitude of its western vertex being at 112.5000° W, 100.0000°° W, and 90.0000°°° W of the Dagger Midline PM. Its eastern vertex is at 86.9966°°° W (Dagger Midline PM) which is very close to 87.0000°°° W. Finally, the latitude of its southern vertex is 8.9429° N of the Arsia Mons Prime Latitude. This is very close to $4\sqrt{5}$ = 8.9443° N (AMPL).

All of this, however, offers only a partial glimpse of the value of this initial square. As it turns out, this square is extremely useful in providing us with the key to the many other squares which also fit the Janssen Crater. It gives us the centre of these squares, and its rotation tells us that the Janssen Crater can be fit to a square whose orientation is synchronized to the cardinal points of the compass. In Chapter 2 we learned that the sizes of squares fitted to craters could be expressed as notes in the chromatic scale. In Table 2.2, this particular square fitting the Janssen Crater was assigned the interval name of M7 which is a major 7th. It was located in the octave containing the reference fundamental square whose diagonal size is 2 latitude degrees. Now it can be seen from Fig. 8.2 that there are a lot of straight line segments in the crater perimeter and other regions in the crater which were not fit by this initial square. Both the rotation and size of the square would have to be altered in order to fit these straight lines and no single square can fit them all. So what I am going to show you next is a series of squares which fit these straight lines. The squares are generally not defined by enough information to be created on their own; they depend on the coordinates of the centre of the first square.

Fitting Other Linear Sections of the Crater Perimeter

The first modified square, shown in Fig. 8.3, was obtained by reducing the rotation of the northwest side of the square to 36° and decreasing the diagonal size to 3.33 latitude degrees. This would make it a major 6th interval in the octave containing a reference square whose diagonal size is

Fig. 8.3: *Square rotated so that its northwest side has a bearing angle of 36°. Its diagonal size is 3.33 latitude degrees. It aligns well with the northwest linear section of the crater perimeter. The square's southeast side aligns to the inside edge of a small linear section in the crater wall near the south vertex. USGS Astrogeology.*

2 latitude degrees. The new square fits the long linear section of the crater perimeter on the northwest side. It also fits a short linear region in the middle of the crater wall near the southern vertex of the square. The centre of the square is the same as for the first square.

Continuing on in the counterclockwise direction, the square was rotated so as to give the northwest side a bearing angle of 30° (Fig. 8.4). This time the size of the square's diagonal was kept to the original 3.75 latitude degrees. Now the square fits a linear section of the crater perimeter on the northeast side. It also gave a good fit to the outer corners of a series of 3 small notches (see inset upper left) in the perimeter of an auxiliary crater embedded in the west side of the Janssen Crater.

In Fig. 8.5, the square has been further rotated in the counterclockwise direction so that the northwest side of the square has a bearing angle of 24°and the size of the diagonal of the square is reduced to 3.60 latitude degrees. The corresponding note on the chromatic scale is a minor 7th in the octave whose fundamental square has a diagonal size of 2 latitude degrees. It can be seen that the latest square creates an excellent fit to the long linear segment of the crater perimeter on the southwest side of the square.

I then noticed a second relatively long linear stretch of the crater perimeter on the southwest side immediately to the east of the one that was just fit. I rotated the current square further in the counterclockwise

Fig. 8.4: *The third square has a diagonal size of 3.75 latitude degrees and is rotated so that its northwest side has a clockwise bearing angle of 30°. This square aligns perfectly with a linear edge on the northeast perimeter of the Janssen Crater. It also aligns to a series of notches (see inset) in the perimeter of a small auxiliary crater on the Janssen Crater's west side. USGS Astrogeology.*

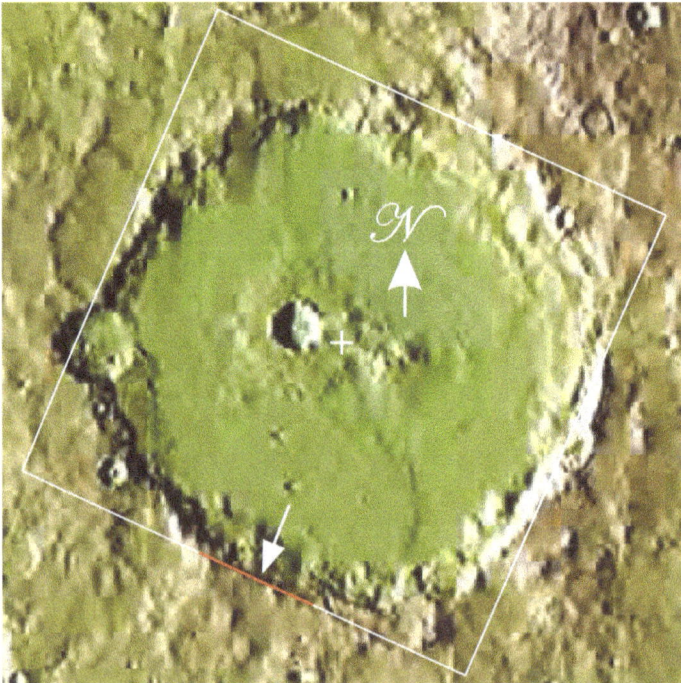

Fig. 8.5: *A square with diagonal size of 3.60 latitude degrees rotated so that its northwest side has a bearing angle of 24°. The square makes a good fit with a linear portion of the crater perimeter on the southwest side of the square (see arrow). Square size corresponds to a minor 7th note (ratio = 9:5) on the chromatic scale. USGS Astrogeology.*

direction until its side became parallel to this new stretch of crater perimeter. This gave the northwest side of the square a bearing angle of 18°. However, the square was too large to fit this other linear section of

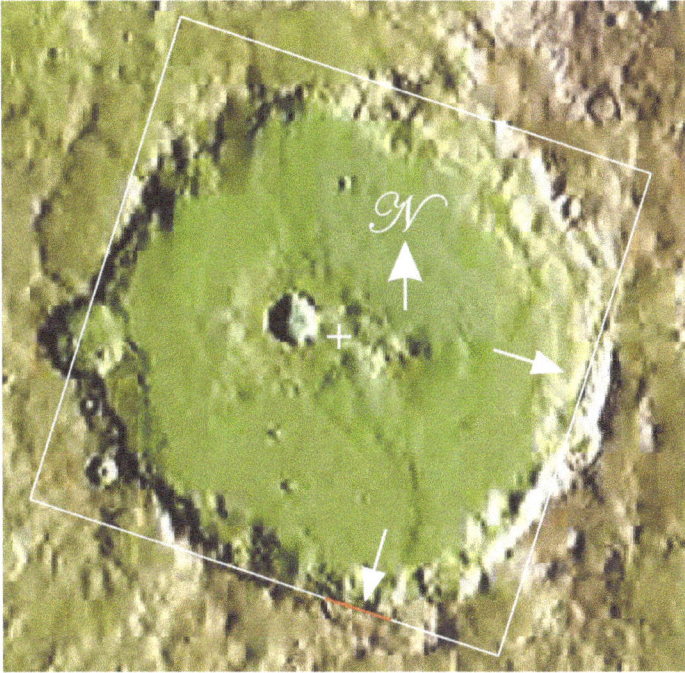

Fig. 8.6: *Square in the previous figure reduced to a diagonal size of 3.5 latitude degrees and rotated so that the northwest side has a clockwise bearing angle of 18°. This square fits a linear section of the crater perimeter just east of that fitted by the square in Fig. 8.5. The size of the square corresponds to a frequency ratio of 7:4 which is an alternate ratio used for the minor 7th note. USGS Astrogeology.*

the crater perimeter until I reduced the diagonal size from 3.6 to 3.5 latitude degrees (red line on square in Fig. 8.6). But this size corresponded to an interval ratio of 7:4 rather than the 5:3 (M6) ratio needed for the next note in the chromatic scale. It turns out that 7:4 produces what is known in music as the harmonic minor 7th and is sometimes used (e.g., in Barbershop Quartet music) to produce a different quality of sound than the more usual ratio of 9:5. So amazingly, the Janssen crater gives us not 1 but 2 versions of the minor 7th note, revealing a level of sophistication that should be expected from such an advanced civilization. The southeast side of the square passes along the middle of a trench in the eastern wall of the crater (see arrow on right).

Instead of going counterclockwise, the next rotation of the square goes clockwise by 15° from our original 45° to give the northwest side of the square a total bearing angle of 60° (Fig. 8.7). Like the square in Fig. 8.3, this square has a diagonal size of 3.33 latitude degrees and would therefore be a major 6th interval in the octave having a reference fundamental of 2 latitude degrees. It shows a good alignment to a linear section of the crater perimeter on the southwest side (lower left arrow, Fig. 8.7). The southwest side also passes through a channel in the crater wall which connects to the auxiliary crater on the west side of the Janssen Crater (upper left arrow, Fig. 8.7). Further to the northwest, the southwest side of the square passes through the centre of a tiny crater inside the western auxiliary crater.

Fig. 8.7: *Fit of a square to the crater perimeter on the southwest side (lower left arrow). The square has a diagonal size of 3.33 latitude degrees and a clockwise bearing angle of 60° for the northwest side. The southwest side also passes through a channel in the Janssen Crater wall (upper left arrow) and through the centre of a tiny crater in the western auxiliary crater. USGS Astrogeology.*

e Square Rotations

There are 2 contiguous linear sections of the crater perimeter on the eastern side which have yet to be fit. The first of these is oriented in a north-south direction so it should line up with a square which has a bearing angle of 0°, i.e., one which has its north and south sides parallel to lines of latitude. However, when I tried to use the original diagonal size of 3.75 latitude degrees, I found that the square was slightly too small. I had to increase it to 3.82 - 3.85 latitude degrees to get a reasonable fit. But there were no integer ratios in that range that made any sense so either Martian music was being creative or else another concept was required. I was mulling over this when it occurred to me that since the latitude of the centre of the squares was equal to e, the diagonal size of this square might also be related somehow to e. I examined the products of the other major irrational numbers used on Mars with e, and it did not take long to discover that the product of $\sqrt{2}$ and e was 3.8442, just the size of number that I was looking for. The fit of a square of this diagonal size in latitude degrees is shown in Fig. 8.8. When I thought about it a bit further, it made even more sense, since a diagonal of this size belonged to a square whose side length was exactly e latitude degrees. Of course! The Janssen Crater celebrated the mystical constant e both in the latitude of its centre and in the side length of one of its fitted squares! However, not just

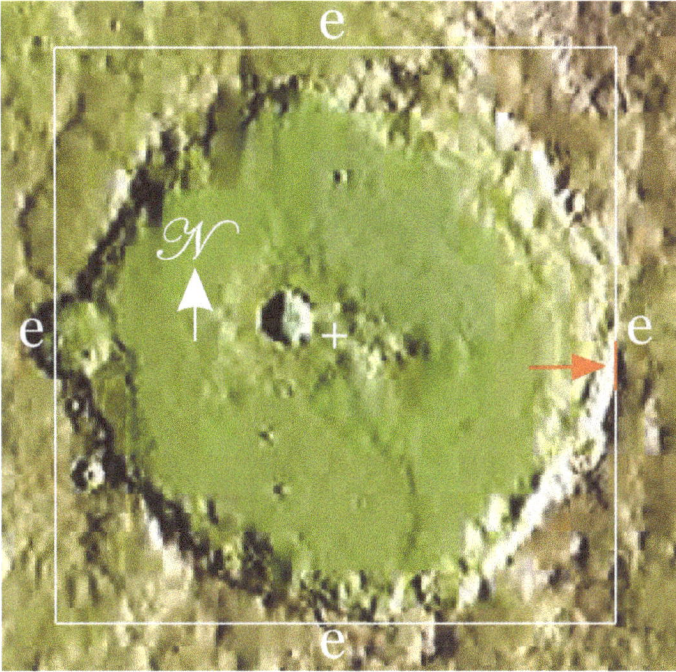

Fig. 8.8: *Square expanded so that its diagonal size equals (√2)e or 3.8442 latitude degrees. The north and south sides are parallel to lines of latitude. Each side of the square is equal to e latitude degrees in length. The square fits a north-south oriented linear section of the eastern crater perimeter (red arrow). USGS Astrogeology.*

one square, but many more as I was to find out. We are now in uncharted territory with "square" craters. All of the previous encounters with "square" craters were with squares sized according to the diagonal fitting a music interval ratio from the chromatic scale. This square is sized according to its side length, not its diagonal.

That left me with the other contiguous linear section of crater perimeter to fit which lies immediately south of the linear section I just fit. By testing several rotations of my latest square, I found that a rotation that gave the square a bearing angle of 18° in the clockwise direction for the northwest side provided the best fit (Fig. 8.9, red arrow). The square also fit the bottom edge of the wall of a small crater near the northern vertex of the square (upper left white arrow). Now I had 2 squares with a side length of e fitting the Janssen Crater.

This is a good place to summarize the findings so far. In Table 8.1, the summary shows that there are 2 examples each for 3 different sizes, and 1 example each for 2 variants of a minor 7th note. What I would like to focus on, though, is the bearing angles. All of the nonzero bearing angles are evenly divisible by 3 and, except for 45°, by 6. This stimulated my thinking again. What would happen if I were to examine squares with other bearing angles which were divisible by 3? But what size should I test? I experimented with several sizes and came to the conclusion that the diagonal size giving the most consistent results was 3.8442 latitude

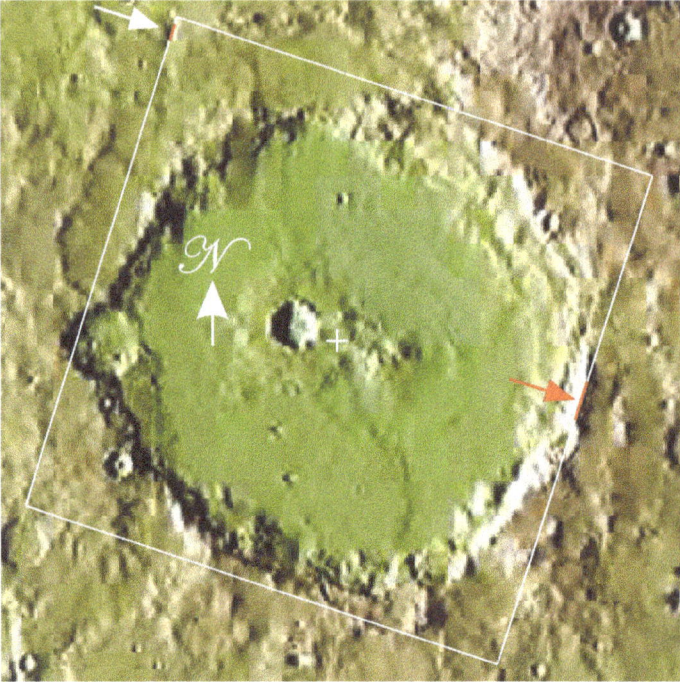

Fig. 8.9: *A square of side length e latitude degrees rotated to a clockwise bearing angle of 18° for its northwest side. The square fits a linear section of the eastern crater perimeter (red arrow) which is contiguous to the linear section in the previous figure. The square also fits the bottom edge of the crater wall of a small crater near the northern vertex of the square (upper left white arrow). USGS Astrogeology.*

degrees. This made very good sense since it would only be appropriate for a crater celebrating the value of e to have lots of squares with a side length of e. I will henceforth refer to them as e squares.

Since it would take too much space to show everything, I will show detailed results for the first 5 rotations in 3° steps, and then I will show all 30 of the steps together in one figure. In the squares that have been shown so far, alignments have been principally made with the perimeter of the Janssen Crater. With the rotations of the e square that I am now going to show you, alignments are often made with structures lying outside of the crater perimeter as well as with the crater perimeter. In Fig. 8.10, the e square with a bearing angle of 3° for its west side is shown. Its east side is in alignment with the eastern edge of the apron of a small crater

Table 8.1: *Summary of squares fitting linear sections of the Janssen Crater perimeter.*

Bearing Angle (°)	Diagonal Size (latitude °)	Interval Ratio	Music Note
60	3.33	5:3	Major 6
45	3.75	15:8	Major 7
36	3.33	5:3	Major 6
30	3.75	15:8	Major 7
24	3.60	9:5	minor 7
18	3.50	7:4	minor 7
18	3.84	N/A	N/A
0	3.84	N/A	N/A

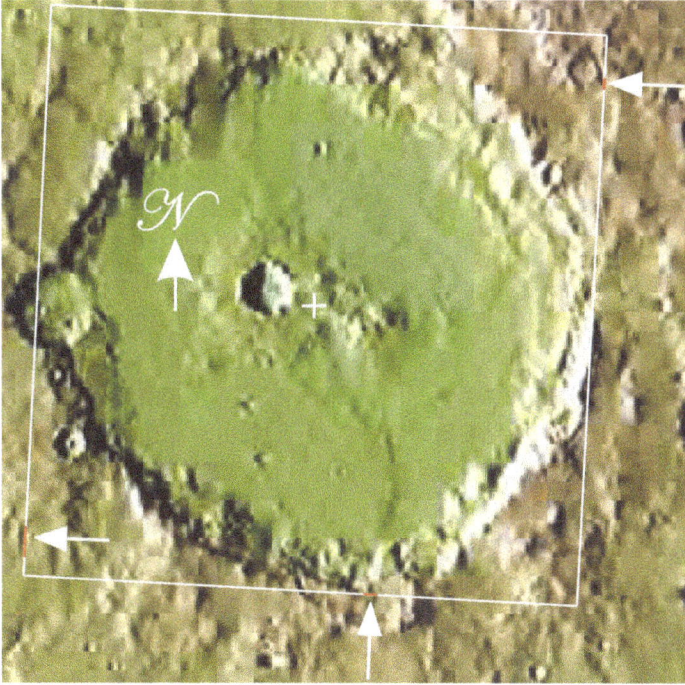

Fig. 8.10: *An e square rotated so that its west side has a bearing angle of 3°. The east side of the square aligns to the eastern edge of the apron of a small crater to the northeast. Alignments with the west and south sides occur with the edges of brown areas (2 arrows at bottom). USGS Astrogeology.*

lying to the northeast of the Janssen Crater. Another alignment is with the southern edge of a very small dark brown structure just beyond the crater perimeter on the south side of the square. A 3rd alignment occurs with the eastern edge of an elongated brown coloured area lying to the southwest of the Janssen Crater.

I next rotated the e square another 3° in the clockwise direction to give the square a bearing angle of 6°. This is shown in Fig. 8.11 in which the east side of the square makes an alignment with a different linear section of the Janssen Crater eastern perimeter. The east side also makes an alignment with the western edge of a long white area lying to the northeast of the Janssen Crater. The west side aligns with the eastern edge of a short brown area lying to the southwest of the Janssen Crater.

The next rotation gives the e square a bearing angle of 9°, and this is shown in Fig. 8.12. The west side of the square is aligned with the western edge of a small crater lying within the auxiliary crater on the west side of the Janssen Crater. To the north of this, there is an alignment of the west side of the square with the eastern edge of a brown coloured area. The east side of the square makes an alignment with the southeast corner of a brown coloured area just beyond the northeast perimeter of the Janssen Crater. The east side is also aligned with the east side of a white/brown coloured structure near the south vertex of the square. This structure is notable in that it is involved in the next 2 rotations of the e square by 3

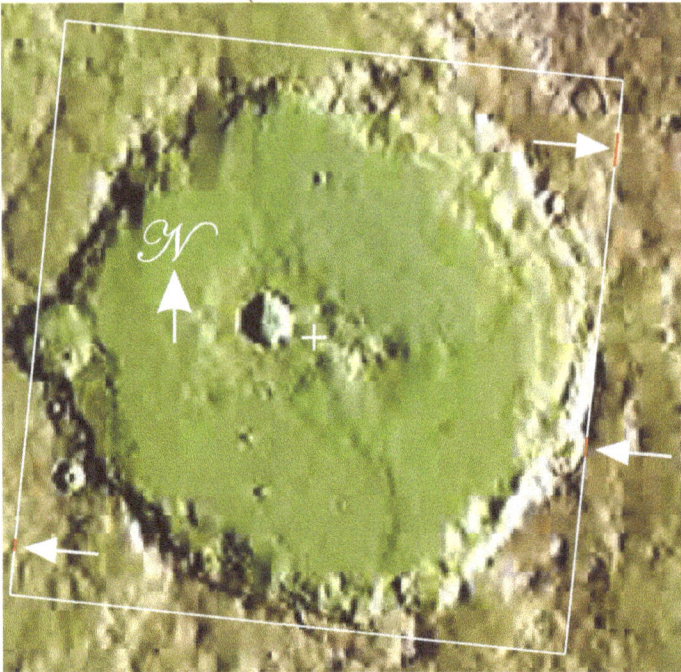

Fig. 8.11: *An e square rotated so that its west side has a bearing angle of 6°. The east side of the square aligns to a linear segment of the eastern perimeter of the Janssen Crater. The east side also aligns with the western edge of a white area near the northeast vertex of the square. The west side aligns with the eastern edge of a small brown area. USGS Astrogeology.*

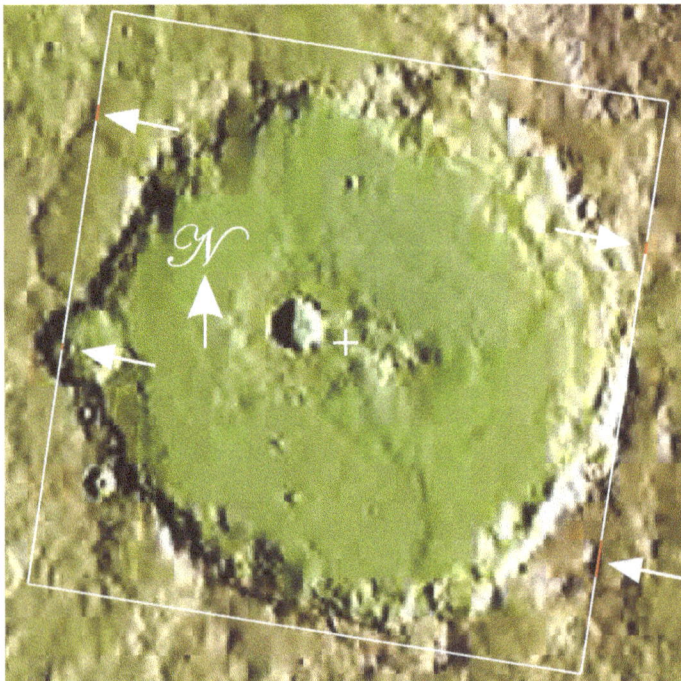

Fig. 8.12: *An e square rotated so that its west side has a bearing angle of 9°. The west side of the square aligns to the western edge of a small crater in the auxiliary crater in the western part of the Janssen Crater. Other alignments occur with the edges of brown coloured areas (see arrows). Note particularly the align-ment pointed to by the lower right arrow. USGS Astrogeology.*

degree steps.

The 4th rotation of the e square is found in Fig. 8.13. Its west side has a bearing angle of 12° and its east side makes an alignment with the

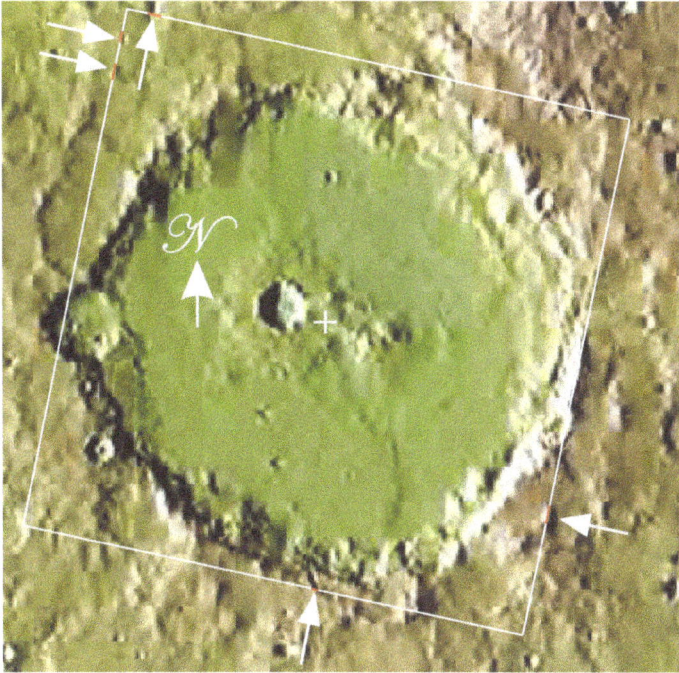

Fig. 8.13: *An e square whose west side is rotated to a bearing angle of 12°. It aligns to the interface between the dark and white areas of the structure near the south vertex (arrow bottom right). It also aligns to the south end of a channel (bottom arrow), to the interface between light and dark areas (lower arrow upper left) and to the edges of 2 small craters (2 top arrows upper left). USGS Astrogeology.*

interface between the white and dark regions of the same structure whose east side aligned with the square in the previous rotation. There is also an alignment of the south side with the southern edge of what appears to be a channel emanating from the southern region of the Janssen Crater. Near the northwest vertex of the square, alignments are made (from lowest arrow in a clockwise direction) with the interface between light and dark regions of a structure, with the west edge of a small crater and with the south edge of another small crater.

The 5th rotation (west side bearing angle = 15°) of the 3 degree steps of the e square is shown in Fig. 8.14. This square aligns with the western edge of the same white and dark structure lying to the southeast of the Janssen Crater which aligned to the previous 2 squares. On the west side of the Janssen Crater, the square aligns with the corner of a right-angled notch in the perimeter of the auxiliary crater. This is shown at higher magnification in an inset at the bottom left of the figure.

With the 2 rotations of 0° and 18° from the first set of data (Figs. 8.8 and 8.9), I have now shown detailed results for 7 successive rotations of the e square. All of these rotations make significant alignments either with structures or landforms lying outside the perimeter of the Janssen Crater, or with the crater perimeter itself. Similar results were obtained with the 23 other rotations which would require too much space to put in individual figures. However, before showing you how they create a

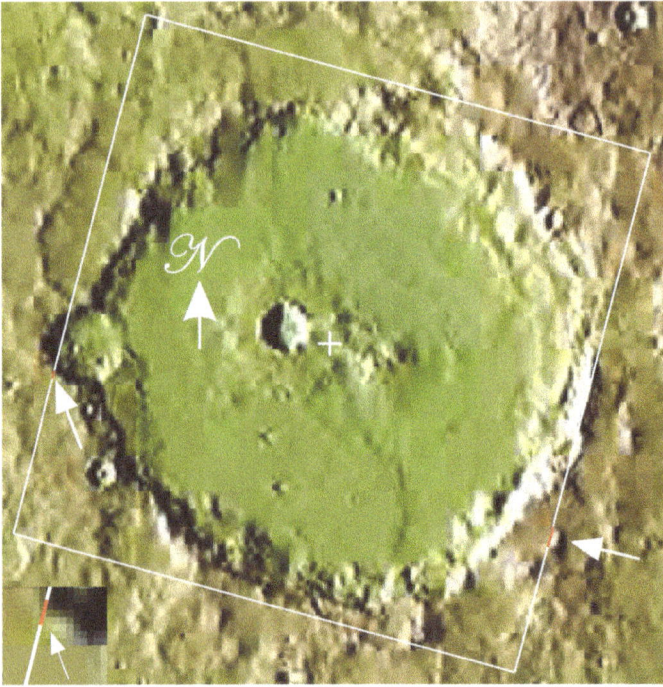

Fig. 8.14: *An e square with its west side rotated to a bearing angle of 15°. The east side of the square aligns with the western edge of the white and dark structure (arrow bottom right) which aligned different parts to the previous 2 rotations. The west side of the square aligns with the corner of a notch in the perimeter of the auxiliary crater (see inset bottom left). USGS Astrogeology.*

picture when put together, I would like to show you just one more rotation of the e square in a separate figure, and that is the 45° rotation. If you remember, the northeast and southeast sides of the 45° rotation of the

Fig. 8.15: *An e square with its northwest side rotated to a clockwise bearing angle of 45°. This square includes the outcroppings of the crater perimeter (see insets top and bottom left) that were not fit by the square with a diagonal size of 3.75 latitude degrees as shown in Fig. 8.2. USGS Astrogeology.*

square with the diagonal size of 3.75 latitude degrees (Fig. 8.2) did not extend quite far enough to include small outcroppings of the crater perimeter. These outcroppings form the alignment points for the e square in Fig. 8.15 and are now very nicely included in the square. An enlargement for the outcroppings on the northeast side is shown in an inset at the upper left, and an enlargement for the outcroppings on the southeast side is shown in an inset at the lower left of the figure.

Janssen's Wheel Revealed

When all 30 of the squares are shown together in Fig. 8.16, the result is a stunning geometric figure of a wheel which encircles the periphery of the Janssen Crater. The centre is located at 37.5495° E 2.7183° N. The inner boundary of the wheel follows much of the crater's perimeter for the eastern half of the circle. However, it lies outside of the crater perimeter for most of the west half of the inner circle. The inner circle passes through the middle of the central peak of the small crater just south of the western auxiliary crater. It also aligns with the western border of a tiny crater lying between the auxiliary crater and the crater with a central peak. The inner diameter of the wheel is equal to e latitude degrees (161.13 km) whereas the outside diameter of the wheel is equal to the e square diagonal size of ($\sqrt{2}$)e latitude degrees (227.87 km). Any point on the inner boundary such as the central peak of the small crater will be e/2 latitude degrees or 80.56 km from the centre of the wheel (indicated by a white cross in Fig. 8.16).

The pattern formed by the 30 e squares gives the appearance of a doughnut or a torus. The division of the circle by squares with bearing angles every 3 degrees might point to a coordinate system of 120 degrees, each equal to 3 regular degrees, for longitude and latitude. This would permit 6 divisions of 5 degrees each for the latitudes of each hemisphere. Interesting latitudes for this system would be 9.00° N for the centre of the outer perimeter of the Sharonov Crater, 5.99° N for the AscSC2 crater, 4.00° N for the Pentagon Pyramid, 12.00° N for Issedon Tholus, and 1.00° S for the 6E3S Crater (see Chapter 10).

The torus shape is found abundantly in nature. It occurs in apples, oranges that have been peeled, hurricanes, and the magnetic fields of the earth, sun and galaxy. In toroidal energy fields, the energy flows in one side, through the central axis, out the other side and then wraps around to the first side once again. We have used the torus shape in human engineering to produce inductors and transformers and also to confine hot plasma (electrons and ions) in nuclear fusion devices. Whether the torus shape created by the Janssen e squares reflects its occurrence in

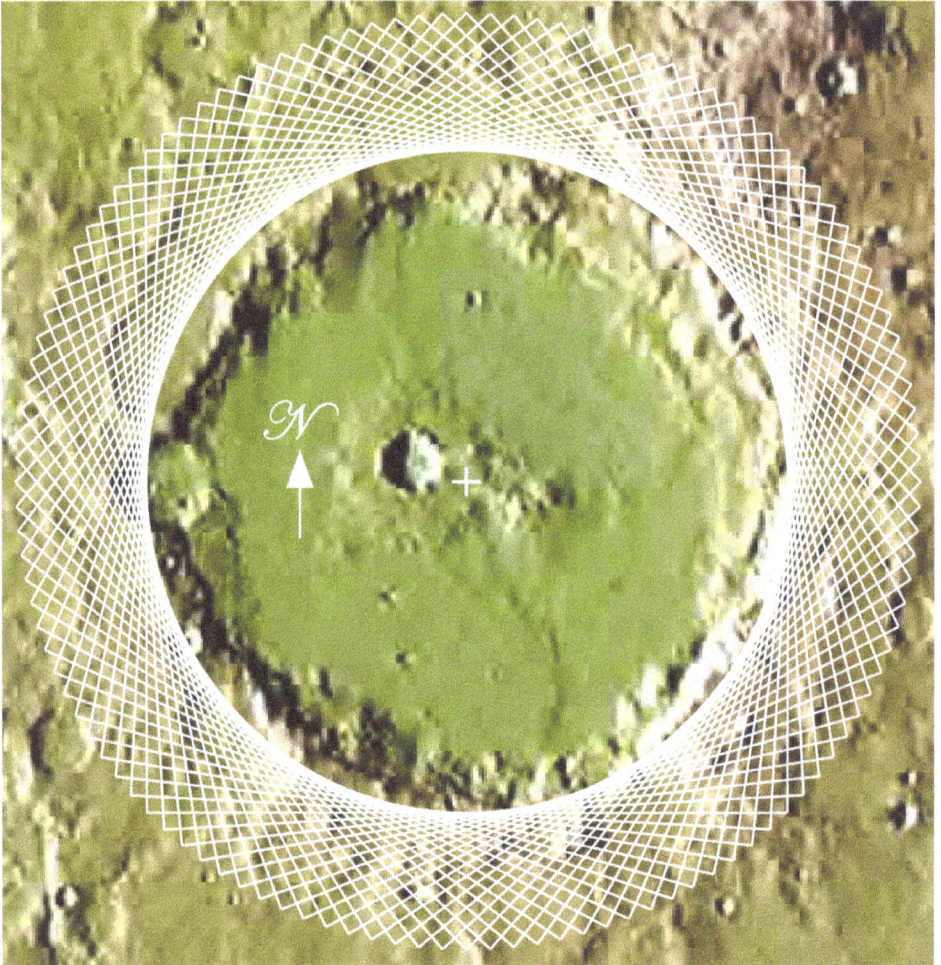

Fig. 8.16: *Janssen's Wheel created by superimposing 30 e squares having bearing angles increasing in steps of 3°. The beautiful figure which results has the appearance of a torus. It may have been used as a compass rose pointing to 120 different wind directions. The centre is located at 37.5495° E 2.7183° N. USGS Astrogeology.*

nature or represents a schematic for a component of an energy system is open to speculation. If the latter is the case, it would appear that this energy system utilizes the number e and perhaps the number 3. Regardless, the use of the value of 3 degrees in this geometric figure is one more instance of the importance the Martians gave to the number 3.

Another possible interpretation of the Janssen Crater wheel is that it could be what is termed a compass rose or design found on compasses to indicate the various directions or winds. The modern compass rose has 8 winds, i.e., the 4 cardinal directions (N, E, S and W) and the directions half way between the cardinal points (NE, SE, SW and NW). More ancient compasses have greater resolution, pointing to 16 and even 32 different wind directions. All of these compasses are not compatible with the Janssen Crater wheel since their division of the circle is based on a factor of 2 instead of 3. There was another compass rose, however, which was called the classical rose. It was used by the ancient Greeks and Romans and had 12 wind directions. It would therefore be compatible with the Janssen Crater wheel but would have one-tenth the resolution, pointing to a separate wind direction every 30 degrees rather than every 3 degrees. In the light of this discussion, the concept of a compass rose would seem to be a very good explanation for the intricate Janssen Crater wheel although one would expect that there would be a special emphasis given to the 4 cardinal points. Perhaps this was accomplished by covert markers which are indicated by white arrows in Fig. 8.17. Here it can be seen that due north from the centre of the squares is marked by the west edge of a small crater in the northern interior of the Janssen Crater. The eastern direction is marked by the centre of a small crater to the east of the Janssen Crater. The southern direction is indicated by the interface between the light and dark regions of a small crater to the south of the Janssen Crater. Finally, the western direction is marked by the middle of the western tab that projects from the auxiliary crater. The black arrow points to a step in the perimeter of a crater lying inside the Janssen Crater.

Fig. 8.17: *The cardinal points of the compass can be obtained by lining up key sites which are aligned with the centre of the squares fitting the Janssen Crater. These are pointed out by the white arrows and are described in the text. The black arrow indicates a step in the perimeter of a crater inside the Janssen Crater which is aligned to the east-west compass directions. USGS Astrogeology.*

This step aligns to the east-west compass points.

Due to the accuracy with which I was able to determine the coordinates of the centre of the Janssen Crater, much has been revealed about the great importance the Martians assigned to the value of the irrational number e. Both the latitude of the centre and the side length of the squares composing Janssen's Wheel are equal to e latitude degrees leading one to wonder if this crater might have been called the 'e' Crater by the Martian architects. Or it may have been called the Natural Log Crater since e is the base of the natural log system. An unnamed crater which will be discussed in Chapter 10 also shows a strong focus on e although this does not seem to be its only purpose nor is it as powerful an indicator of e as the Janssen Crater.

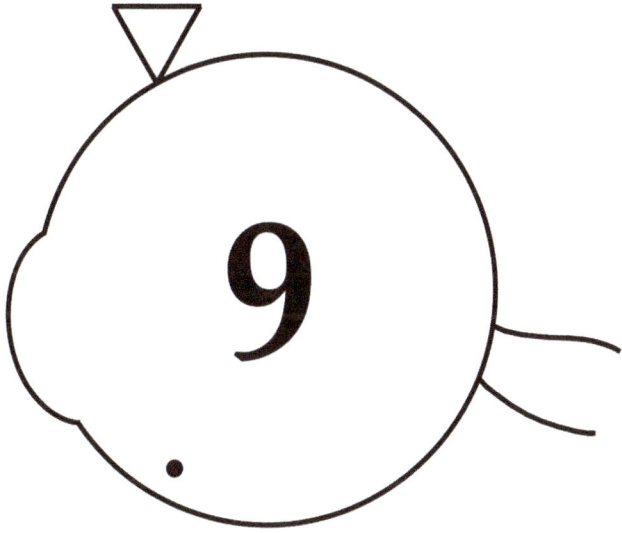

The Elorza Alignments

Approximately 30 degrees due south and slightly to the east of the Sharonov Crater lies an interesting site called the Elorza Crater (Fig. 9.1). The crater is named after a small town in Venezuela. It is located in the southern hemisphere just below the equator and the

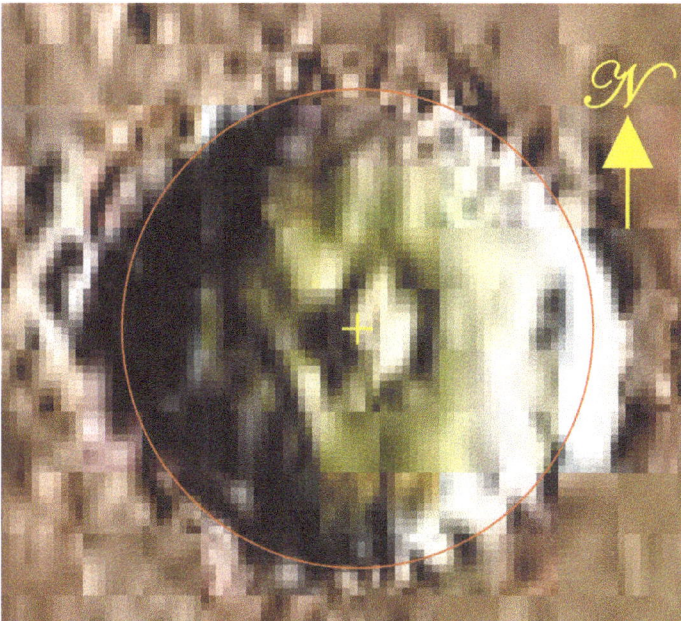

Fig. 9.1: *The perimeter of the Elorza Crater fits a circle well on the north and south sides but falls short of the crater perimeter on the east and the west sides. Coordinates of the centre of the red circle (yellow cross) are 304.7904° E 8.7500° S. USGS Astrogeology.*

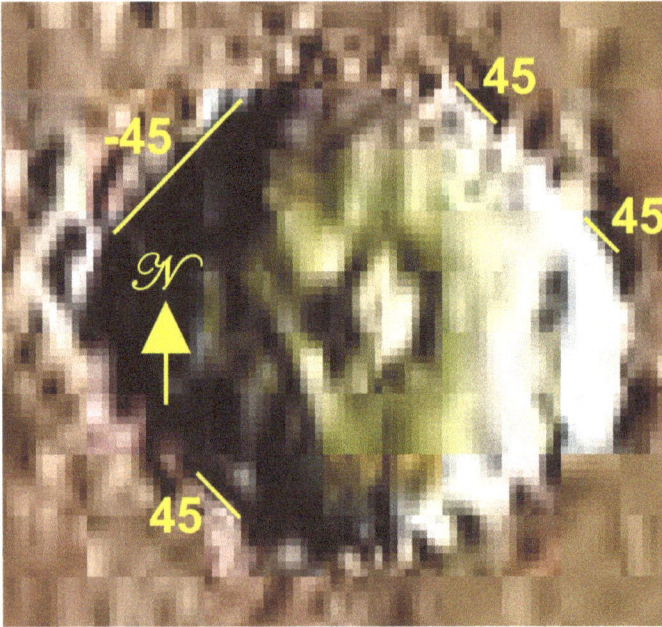

Fig. 9.2: *Straight line segments with bearing angles of 45° in the clockwise or counterclockwise directions can be fit to the perimeter of the Elorza Crater. USGS Astrogeology.*

coordinates of the centre of a fitted circle are 304.7904° E 8.7500° S. The diameter of this crater is about 43 km making it slightly less than half the size of the Sharonov Crater. At first glance the Elorza Crater looks round, but looks can be deceiving. The perimeter exceeds the limits of a fitted circle on its east and west sides (Fig. 9.1). When we examine the crater closely, straight line segments are found to occur on the northeast, northwest and southwest perimeter sides and have bearing angles of 45° in the clockwise or counterclockwise direction (Fig. 9.2). This suggests that the Elorza Crater is really a "square" crater in disguise having the most common bearing angle of 45° for its northwest side. To determine the centre and size of a square to fit the crater, however, was not an easy task. I started by trying to fit a square (with a 45° bearing angle for the northwest side) to the farthest outreaches of the crater. This included the outer limits of channels coming off the northwest and southwest sides of the crater, the outer edge of the crater apron on the northeast side, and an outward notch in the crater perimeter on the southeast side. I discovered that a square with a diagonal size of 1.2000 latitude degrees would fit all of these features (see Fig. 9.3). This was the square that I reported in Tables 2.1 and 2.2 for the Elorza Crater in the chapter on square craters. Its size makes it a minor 3rd interval with reference to a square whose diagonal size is 1.0000 latitude degrees. The square has the same centre as the circle in Fig. 9.1. The centre and both east and west vertices have an interesting latitude of 7.0000°°° S and the south vertex is 1.0003°°° S of the

Fig. 9.3: *Square with a diagonal size of 1.2 latitude degrees and a bearing angle of 45° for its northwest side. The square fits the crater apron and outside edges of channels from the main crater on the northwest and southwest sides. Its northeast side fits the outside edge of the crater apron and the southeast side fits the crater perimeter in one location. USGS Astrogeology.*

Arsia Mons Prime Latitude. The centre and both north and south vertices of the square have the very meaningful longitude of $\sqrt{5}\varphi = 3.6180°$ E (Sharonov Tower PM). The west vertex has a longitude of 157.0786° E and 125.6629°°° E (Crater Edge PM). The numerical values of these longitudes are very close to $50\pi = 157.0796$ and $40\pi = 125.6637$.

If the diagonal is reduced to 0.9375 latitude degrees, the square fits linear segments of the perimeter on all 4 sides (Fig. 9.4). The longest linear lengths of crater perimeter which fit the square are on the northeast side. The west vertex of this square is located at 3.1437° E of the Sharonov Tower PM which has a numerical value very close to $\pi = 3.1416$. The north vertex has a latitude of 0.1614°° S (AMPL) which is very close to $\varphi/10 = 0.1618°°$ S (AMPL). The south vertex has a latitude of 1.1191° S (AMPL) which is very close to $\sqrt{5}/2 = 1.1180°$ S (AMPL). This size of square would be a major 7th interval larger than a reference square whose diagonal size is 0.5000 latitude degrees.

The analysis provided so far of the Elorza Crater, while very interesting, by itself does not give this crater enough status to merit a whole chapter devoted to it. It is not until you measure distances and bearing angles to other sites that you start to realize the special properties of this crater. It turns out that the coordinates of the centre of the Elorza Crater are rather unique, placing it at an alignment node with respect to several of the major sites on Mars. It creates 10 isosceles triangles with

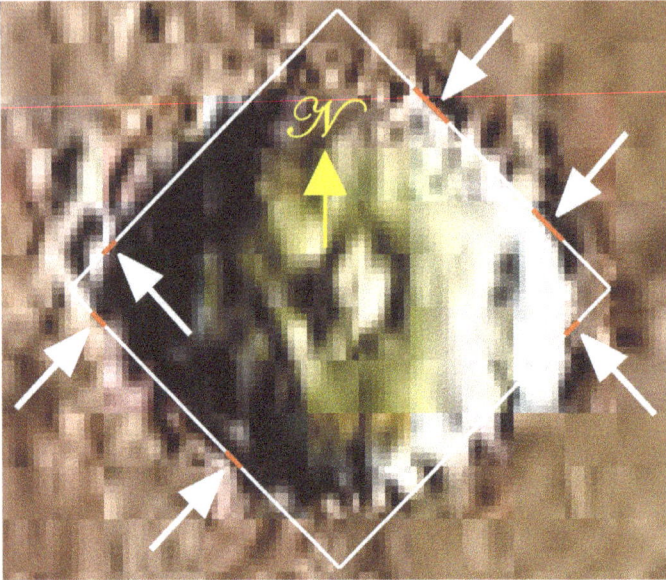

Fig. 9.4: *Square with a diagonal size of 0.9375 latitude degrees and a bearing angle of 45° for the northwest side. The square fits short linear edges of the crater perimeter on all 4 sides. USGS Astrogeology.*

other sites, 4 of which have their equal sides within 1 km of each other and the other 6 have their equal sides within 6 km of each other (Table 9.1). It creates 2 of these isosceles triangles with aspects of the Sharonov Crater (see Fig. 9.5 for identification). Thus the Elorza Crater is almost equidistant from both the centre of the circle which fits the outer perimeter of the Sharonov Crater (white X in Fig. 9.5) and the centre of the S2 crater inside the Sharonov Crater. This distance is close to 5/8 of the equatorial radius (R) of Mars. Also the distance from the Elorza Crater to the tip of the triangle which touches the outer perimeter of the Sharonov Crater is within 1 km of the distance from the Elorza Crater to the S3 crater which lies just outside of the Sharonov Crater to the northeast. This distance is close to the sacred distance of $\pi\sqrt{5}R/11$ km. Several other sites create isosceles triangles with the Elorza Crater. The Paros Crater is approximately the same distance from the Elorza Crater as the AscSC1 Crater with a value close to the sacred distance of $9R/(\pi^2)$ km. The Pavonis Mons Caldera, Issedon Tholus and the Ayacucho Crater are at a distance close to $\pi\varphi R/5$ km, a distance formula which strongly references the pentagram which these 3 sites pay homage to. Jovis Tholus and Ulysses Tholus are almost exactly equidistant from the Elorza Crater in 2 ways. The first way is the match between the distances between the eastern centre of Jovis Tholus and the centre of the Ulysses Tholus Caldera with a distance close to $e\sqrt{3}R'/4$ km (R' is the northern polar radius of Mars) from the Elorza Crater. The second way is the distance of approximately $6R'/(\varphi\pi)$ km from the Elorza Crater to the western centre

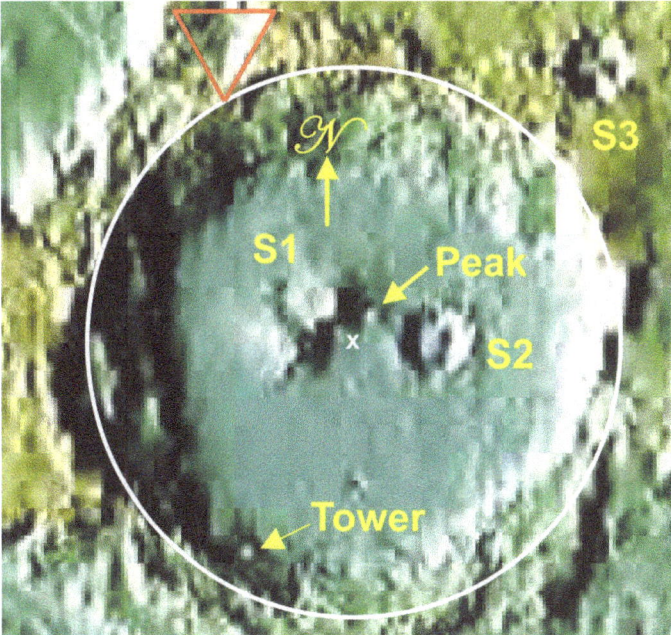

Fig. 9.5: *The Sharonov Crater with identification of various sites which are associated with it. The S2 and S3 sites are craters. The white X marks the centre of the circle fitted to the crater perimeter. The place where the red triangle touches the white circle marks the location of the Sharonov Triangle tip and the Sharonov Triangle PM. USGS Astrogeology.*

Table 9.1: *Distances from the Elorza Crater centre to other sites.*

Site		Distance (km)		
		Sacred	Actual	Diff.
Sharonov Crater circle centre	5/8R	2122.62	2127.14	4.52
Sharonov S2 crater centre	5/8R	2122.62	2126.27	3.65
Sharonov Triangle tip	$\pi\sqrt{5}R/11$	2168.87	2170.67	1.80
Sharonov S3 crater centre	$\pi\sqrt{5}R/11$	2168.87	2171.46	2.59
Paros Crater	$9R/\pi^2$	3096.95	3090.38	-6.58
AscSC1 Crater	$9R/\pi^2$	3096.95	3096.07	-0.88
Pavonis Mons Caldera	$\pi\varphi R/5$	3452.71	3446.73	-5.97
Issedon Tholus	$\pi\varphi R/5$	3452.71	3452.16	-0.55
Ayacucho Crater	$\pi\varphi R/5$	3452.71	3450.79	-1.92
Jovis Tholus East	$e\sqrt{3}R'/4$	3973.96	3972.19	-1.77
Ulysses Tholus Caldera	$e\sqrt{3}R'/4$	3973.96	3972.17	-1.79
Ulysses Tholus N Crater	$6R'/(\varphi\pi)$	3985.13	3978.73	-6.40
Jovis Tholus West	$6R'/(\varphi\pi)$	3985.13	3981.31	-3.81
Ulysses Tholus	$6R'/(\varphi\pi)$	3985.13	3981.83	-3.29
Sharonov Tower	$e^2R/12$	2091.22	2093.12	1.90
Ascraeus Mons	$e^2R/8$	3136.83	3137.69	0.86
AscSC1 Crater	$9R/\pi^2$	3096.95	3096.07	-0.88
AscSC1a Crater	$8R/\pi^2$	2752.85	2752.28	-0.57

of Jovis Tholus, to the survey centre of Ulysses Tholus and to the crater on the north side of Ulysses Tholus.

Also shown in Table 9.1 are distances to sites from the Elorza Crater which are matched in form but not in value (see items below dashed line). The distance to the Sharonov Tower is almost equal to $e^2R/12$ km which is similar in form to the distance of $e^2R/8$ km to the survey centre of Ascraeus Mons. The distance to the AscSC1 Crater is very close to $9R/\pi^2$ km which is similar in form to the distance of $8R/\pi^2$ km to the AscSC1a Crater. The ratio of the sacred distances in these pairs of sites is a music interval in the chromatic scale. Thus the ratio of the sacred distance to Ascraeus Mons to the sacred distance to the Sharonov Tower is $3137/2091= 1.500$ which is the interval for a perfect 5th. The ratio of the sacred distance to the AscSC1 Crater to the sacred distance to the AscSC1a Crater is $3097/2753= 1.125$ which is the interval for a major 2nd.

It is time now to examine the biggest reason why the Elorza Crater merits our attention. The alignment properties of the Elorza Crater do not stop at being equidistant to several major sites. It also lines several sites up with the same bearing angle (all in the counterclockwise direction) from the Elorza Crater. Incredibly, there are no less than 5 such alignments. Firstly, there is the alignment of Uranius Mons with Alba Mons having an average bearing angle of 46.69° (Table 9.2). Next there is the alignment of the northern centre of Tharsis Tholus with the AscSC1 crater having an average bearing of 57.72°. Thirdly, there is the alignment of the survey centre of Pavonis Mons with an amazing 3 other major sites: the caldera of Ulysses Tholus, the Pettit Crater and Albor Tholus having an average bearing angle of 79.85°. For the next alignment, it should be noted that the Biblis Tholus Caldera is elongated in approximately the north-south direction and therefore has to be fit with 2 overlapping circles, one for the northern perimeter and one for the southern perimeter. The bearing angle of the centre of the southern perimeter aligns with that for the Pentagram Pyramid and the Pavonis Mons Caldera at an average angle of 80.80°. The final group of sites to have a common bearing angle with Elorza contains Arsia Mons, the peak in the dagger handle which marks one of the prime meridians, and Apollinaris Mons, with an average bearing angle of 89.41°.

The exact meaning of such alignments to the Martian architects is of course open to speculation. But by using an analogy to astrology, if the Elorza Crater was the Earth, and the other major sites were planets, the presence of so many planetary alignments would certainly be regarded as being highly significant. The alignments are shown diagrammatically in Fig. 9.6 to illustrate how the sites are placed in relationship to each other. The sites in alignment would be said to be in conjunction with each other. Conjunctions enhance the energies of the planets and this can be

Table 9.2: *Groups of sites having a common counterclockwise bearing angle from the centre of the Elorza Crater.*

Site	Bearing (degrees)	Mean Bearing
Uranius Mons	46.59	
Alba Mons	46.79	46.69
Tharsis Tholus North	57.90	
AscSC1 Crater	57.55	57.72
Pavonis Mons	79.80	
Ulysses Tholus Caldera	79.95	
Pettit Crater	79.89	
Albor Tholus	79.75	79.85
Pavonis Mons Caldera	80.83	
Biblis Tholus Caldera S Centre	80.80	
Pentagram Pyramid	80.77	80.80
Arsia Mons	89.43	
Apollinaris Mons	89.40	
Dagger Peak	89.41	89.41

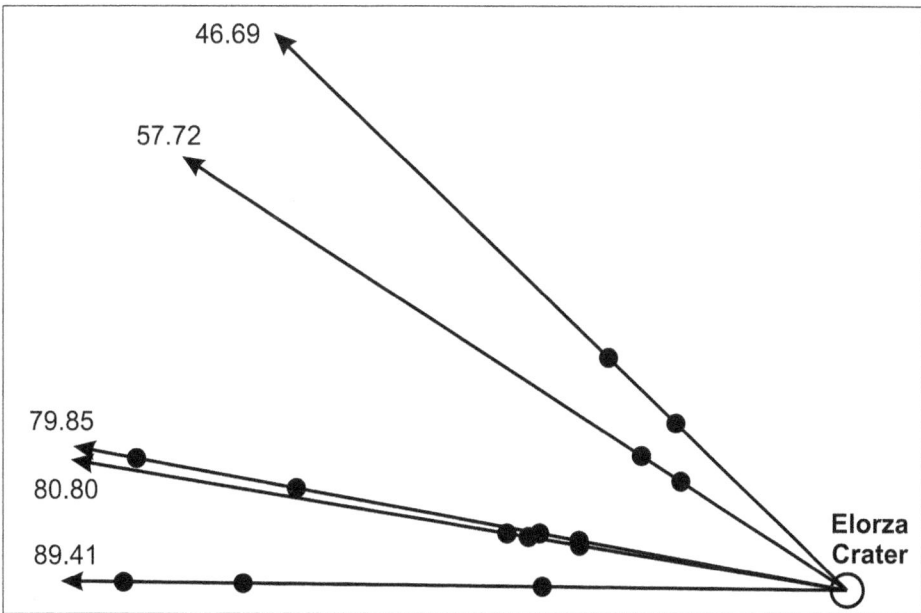

Fig. 9.6: *Diagrammatic representation of alignment of sites (shown by black circles) to the Elorza Crater. Numbers are the average counterclockwise bearing angles of aligned sites in Table 9.2 placed at their relative distances from the Elorza Crater.*

favourable (e.g., a Moon-Venus conjunction) or can instead create stress (e.g., a Saturn-Mars conjunction). The Elorza Crater would therefore probably be regarded as a nodal point of very high energy. Furthermore, the alignments along similar bearing lines are reminiscent of sacred temples and megalithic structures being aligned to straight lines here on Earth. These lines have been called ley lines and are thought by some to be associated with special psychic or spiritual energy.

The Elorza Crater As An Eye

The strong ties of the Elorza Crater to the Sharonov Crater suggest that these 2 sites are somehow supposed to function as a team. Indeed, as we shall find out, both of these craters "view" very distant sites but in different ways. The Elorza Crater is another example of severe inattentional blindness on my part since it took me many years before I finally saw the obvious. The Elorza Crater has the shape of an eye! If one draws an arc of a circle to fit the western perimeter of the crater and joins this to the circle which is shown in Fig. 9.1, you end up with a perfect

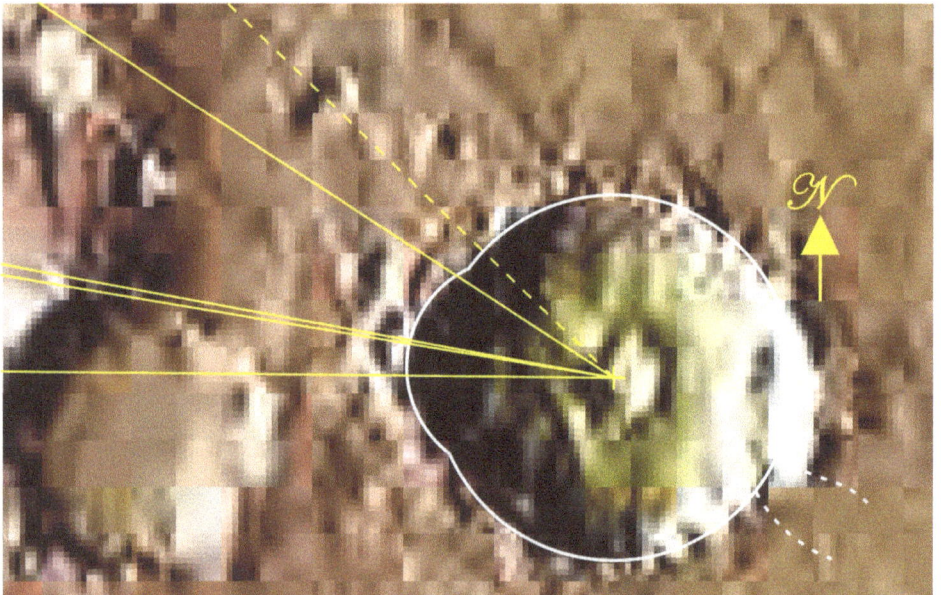

Fig. 9.7: *Outline revealing the fact that the Elorza Crater has the unmistakable shape of an eye. The yellow rays are the lines of alignment emanating from the Elorza Crater as shown in Fig. 9.6. The dashed yellow line goes to Uranius Mons and Alba Mons which are out of the "field of view" of the eye of Elorza. The pair of dashed curved lines at the bottom right of the eye represent the approximate location of where the optic nerve would exit if this was a depiction of the human eye. USGS Astrogeology.*

representation of an eye, complete with its cornea (Fig. 9.7). It is proportioned such that it falls within the range of human eye variation. There does not seem to be a representation of the optic nerve, however, as the approximate location of the nerve shown by dashed white lines in Fig. 9.7 does not correspond to any particular crater feature.

If we assume the focal point of the eye corresponds to the centre of the main circle fitting the crater perimeter (which is the same as the centre of the squares above), the lines of alignment emanating from the Elorza Crater as shown in Fig. 9.6 pass through the cornea of the eye except for the dashed yellow line which goes to Uranius Mons and Alba Mons. These 2 sites would appear to be "out of view" of the Elorza Crater. Hence, even though the focal point chosen in Fig. 9.7 should actually be at the back of the eye to be anatomically correct, moving the focal point towards the rear would put more sites "out of view". The only way to have all sites "in view" would be to move the focal point closer to the front of the eye, but there is no structure inside the crater that would seem to be designed for this purpose. Moving the focal point all the way to the front of the cornea would result in poorer fits to bearing angles and sacred distances in most cases. This suggests that the focal point in Fig. 9.7 was probably the one intended by the architects and that for some reason, the architects did not want Alba Mons and Uranius Mons to be "seen" by the eye. They could have rotated the eye clockwise by about 10 degrees to put these sites in "view", but perhaps they were considered to be too sacred to allow this to happen. Besides, there may be other aligned sites in the southern hemisphere which they wanted the eye to "see".

None of the sites which are in the "view" of the eye could ever be seen by a person standing at the focal point, even with a very powerful telescope. This is because the distances are so great that sites are hidden by the curvature of the planet. Hence, the "viewing" function of the eye is symbolic only.

The fact that the Elorza "eye" is essentially the shape of a human eye has major implications for us here on Earth. The Martians could be our own ancestors in some way, either by direct ancestry from natural breeding or by biological ancestry achieved through genetic engineering. Or perhaps we share a common ancestry from a very ancient civilization elsewhere in the universe. The giant bisected isosceles triangle shaped by Olympus Mons and the Tharsis Montes was postulated in *Intelligent Mars I* to be a representation of the Martian body shape in which the height of the body is equal to the distance covered by the extended arms. This is exactly similar to the human body as shown in the Vitruvian Man drawing by Leonardo Da Vinci. Another example of Martian similarity to human anatomy will be shown with the Sharonov Crater in Chapter 11.

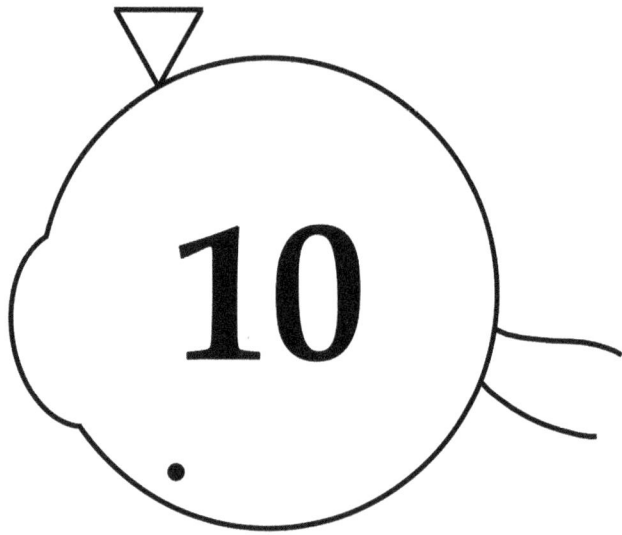

The 6E3S Crater

Not too far from the Elorza Crater, situated about 400 km to the northeast, is a very strangely shaped unnamed crater (Fig. 10.1). It would be hard to imagine what type of meteorite could have created the many notches and straight line segments in its perimeter. Its northwest and northeast edges form an obtuse angle. Its western edge has several steps and notches plus a bulge extending outward in its northern half. The southern perimeter has a long straight line segment with 3 small light-coloured dots superimposed. Its eastern edge has a white coloured platform extending outwards to the east. Regardless of its origins, however, a huge force would have been required to produce such a crater since its diameter is about 83 km. What really attracted my interest though were the coordinates that are found in the central region of this crater when you use the tip of the Sharonov Triangle as the Prime Meridian, namely 6°° E 3° S. Initially these were just theoretical coordinates since the centre of this crater is extremely difficult, if not impossible, to measure by fitting a circle to its perimeter. Due to its irregular shape, one has to make difficult assumptions as to what parts of the crater perimeter are to be used. After many attempts to do so, I came to the conclusion that any estimate of the crater's "true" centre would be biased. However, I noticed that my theoretical coordinates were in an area of brighter pixels which might represent the central peak of the crater. The theoretical coordinates of 6°° E 3° S translate into the conventional coordinates of 307.8547° E

Fig. 10.1: *An unnamed oddly shaped crater, with many notches and straight line segments in its perimeter, is located at 307.8547° E 3.0000° S which are my theoretical coordinates for the centre of this crater. Its western edge has a bulge extending outward. The southern edge has a straight line segment with 3 small light spots on it. USGS Astrogeology.*

3.0000° S. The use of regular degrees for the latitude coordinate and big degrees for the longitude coordinate is reminiscent of the Pentagon Pyramid east-west midpoint with coordinates 12°° E 12° N (Dagger Peak PM). It is possible that it may have been a normal practice for the Martian civilization to simultaneously use both degree systems for navigational as well as architectural purposes.

You may remember from *Intelligent Mars I* that the numbers 3 and 6 are the even divisors of the bearing angles of the vast majority of straight line segments found virtually everywhere on Mars. This crater is no stranger to straight line segments, either within its interior, or as parts of its perimeter. I was able to measure the bearing angles of 12 straight line segments within the crater and a further 13 segments forming parts of the perimeter (Fig 10.2). When you set aside the segments with a bearing angle of 0°, all the remaining segments are evenly divisible by 3, and most of them are also divisible by 6. The most frequently occurring angles are 0°, 45° and 36° as was found for straight line features occurring on the major mountains (see *Intelligent Mars I*). Here there are 9 instances of 0°, 5 instances of ±45° and 4 instances of ±36°. With lines having bearing angles of 0°, ±45° and 90°, we have all of the lines with bearing angles that point to the directions of the 8 most important points on a compass. This suggests that a primary purpose of this crater might be to provide directional information to spacecraft flying overhead. The 4 instances of ±36° and 1 instance of 18° imply a reference to the pentagram since 36° is

Fig 10.2: *A total of 25 straight line segments can be found in the perimeter and interior of the odd-shaped crater. Bearing angles were determined for 13 linear segments in the perimeter and 12 in the interior. There are 2 X-shapes within the crater, one of which is created by the intersection of a curved and a straight line. Negative numbers are bearing angles in the clockwise direction. USGS Astrogeology.*

the angle of the star points and 18° is one-half of this value.

The bearing angle of the northwest side caught my attention since 60° in the clockwise direction is what you would expect if the northwest side of a north-south oriented hexagon were fit to that section of the crater. I experimented with the hexagon shape and found that a good fit could indeed be obtained with a hexagon which had a side-to-side width of 1.3333 latitude degrees (Fig. 10.3). This size would make it a perfect 4th interval above a fundamental width of 1.0000 latitude degrees. The northwest side of the hexagon fit the parts of the crater perimeter which were to the inside, and formed a straight line over a long distance. Except for the west and southwest sides, the other sides of the hexagon fit short sections of the crater perimeter (see arrows). The west side is aligned to the east side of the southern part of a white coloured structure inside the crater interior. The southwest side passes through the centre of the western bright-coloured dot on the southern perimeter of the crater. Most importantly, this excellent fit was obtained when I set the coordinates of the centre of the hexagon to exactly 6°° E 3° S (Sharonov Triangle PM), the very coordinates that I was unable to verify previously by trying to fit a circle to the crater perimeter. Since there is no official name for this crater I decided to call it the 6E3S Crater to refer to its special coordinates.

The longitude of the centre of the fitted hexagon is also 142.0000°° E (Dagger Peak PM). Much more importantly, it is 54.0000°° E of the Pavonis Mons PM and 54/10 = 5.4000°°° E of the Sharonov Triangle PM).

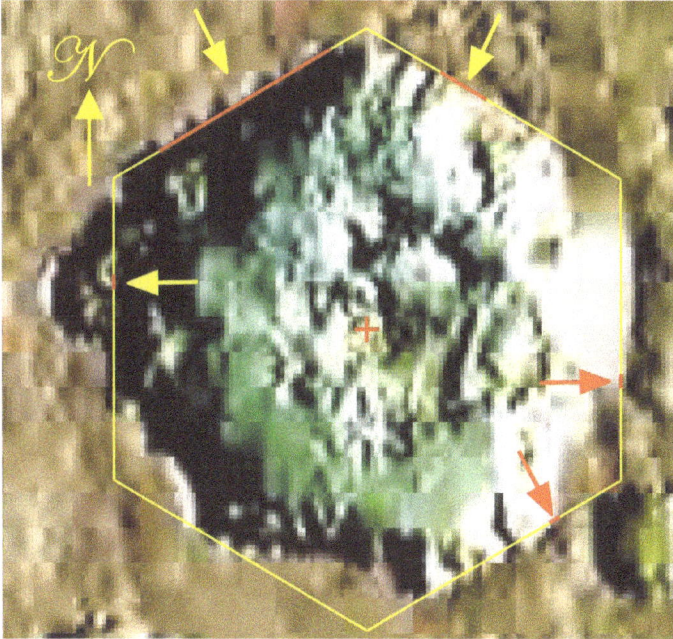

Fig 10.3: *A hexagon with a side-to-side width of 1.3333 latitude degrees and oriented in the north-south direction is well fit to sections of the crater perimeter, especially on the northwest side. The west side is aligned to the southern east edge of a white structure in the crater interior. The coordinates of the centre of the hexagon are 6°° E (Sharonov Triangle PM) and 3° S. USGS Astrogeology.*

These latter values signify the pentagram since 54 is one-half the number of regular degrees between the star points of a pentagram. Furthermore, this longitude is 48.5458°°° E of the Pavonis Caldera PM with a numerical value close to 30φ = 48.5410. Therefore, it would seem that an important purpose of the 6E3S Crater is to honour the pentagram. This is borne out by the relationship of the bright-coloured dots in the south part of the crater to the pentagram (see below). The latitude of the northern ends of the east and west sides of the hexagon is 2.6151° S, the numerical value of which is close to $\varphi^2 = 2.6180$.

Parts of the crater perimeter lay outside of the fitted hexagon. This suggested to me that perhaps there was a slightly larger hexagon which would fit the furthest outlying extensions of the crater perimeter other than its western bulge. I found that by keeping the same centre and increasing the width of the hexagon to 1.4063 latitude degrees, an excellent fit to these extensions was achieved (Fig. 10.4). The size of this hexagon corresponds to one of the integer ratios (45:32) used for the tritone interval above a fundamental size of 1.0000 latitude degree. The expanded hexagon fit the long linear section of the crater perimeter on the east side. It also fit several outward projections of the crater perimeter on the northwest side. The west side fit the western edge of the same white structure whose southern east side aligned to the smaller hexagon. The southeast side fit the furthest outward extension of the crater perimeter. Other than the coordinates of its centre, the most noteworthy coordinate of

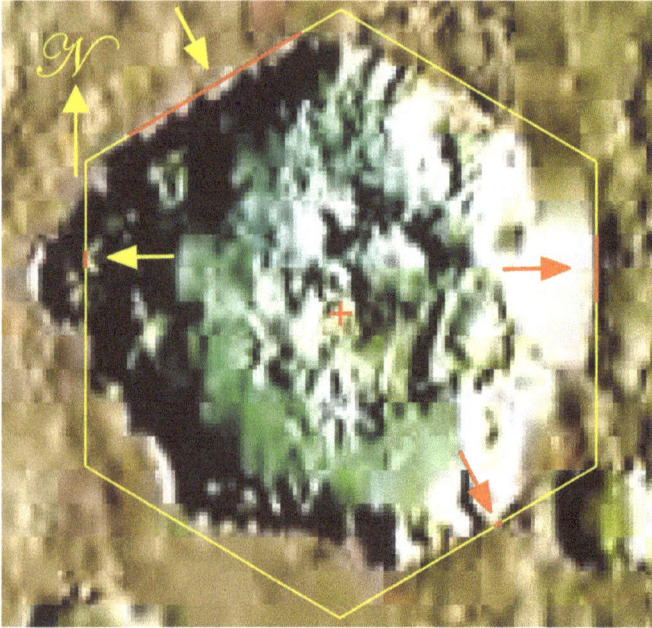

Fig 10.4: *The hexagon in Fig. 10.3 expanded to a width of 1.4063 latitude degrees fits the linear eastern edge of the crater perimeter. It also fit the edges of the outward extensions of the crater perimeter on the northwest side. The west side fits the western edge of the white structure in the crater interior. The southeast side fits the furthest outward extension of the crater perimeter. USGS Astrogeology.*

this hexagon is the average longitude of its east side, being 7.3864° E (Sharonov Tower PM) with a numerical value close to $e^2 = 7.3891$.

Despite the good fits of the 2 previous hexagons to the 6E3S Crater, I found that the latitude of the SE and SW vertices of the smaller hexagon was 2.7079°°° S, and for the larger hexagon was 2.7248°°° S. This suggested to me that there should be a hexagon size where these vertices would be located at exactly 2.7183°°° S, the value of the irrational number e. I experimented with this until I found that the size that would give this latitude for the vertices was a hexagon whose width was 1.375 latitude degrees. Actually the latitude was 2.7175°°° S which is only 3 seconds of a degree smaller so I figured I had the correct size. But would it fit the crater in a meaningful way? The answer can be seen in Fig. 10.5 where this hexagon is shown to fit 2 linear sections of the eastern edge of the crater perimeter. At the west side of the crater, the hexagon fits a linear section of the crater perimeter which is oriented in an exact north-south direction like the linear sections of the east side. The northwest side of the hexagon is aligned with a short linear section of the crater perimeter near the northwest vertex. The southwest and northeast sides align with the furthest extensions of the crater perimeter and the southeast side aligns with the crater perimeter at 2 locations. So this hexagon gives an extremely good fit to features of the crater on all sides. There is one anomaly to this, however. The size of the hexagon corresponds to an interval of 11/8 which is not one of the intervals to be found in the

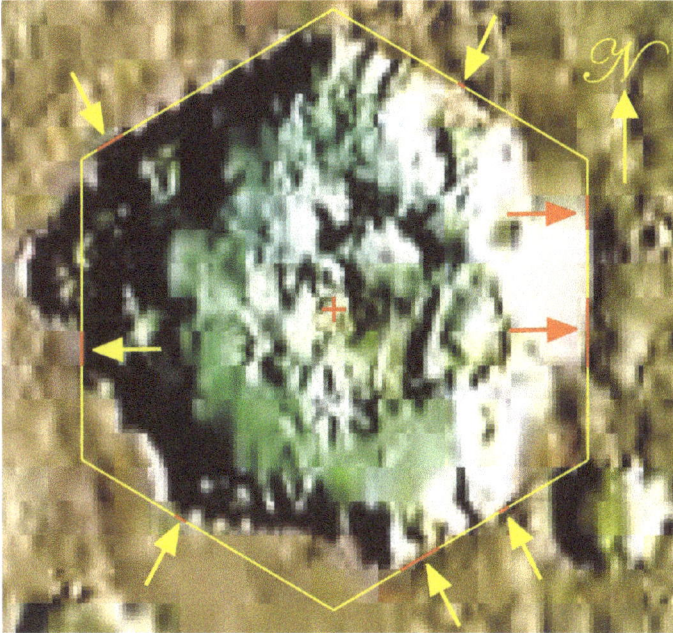

Fig 10.5: *A hexagon with a width of 1.375 latitude degrees fits 2 linear sections of the eastern crater perimeter. It also fits linear sections of the perimeter on the southeast, west and northwest sides. The northeast and southwest sides fit the furthest outward extensions of the crater perimeter. The latitude of the south ends of the east and west sides is e°°° S. USGS Astrogeology.*

standard chromatic scale. It does exists as a music interval in certain types of music under the name of the undecimal tritone or the 11th harmonic interval, but it sits about halfway between the perfect 4th and the tritone intervals. So this is the first deviation from the chromatic scale intervals that I came across for the hexagon sizes and it reveals a greater complexity than I expected. Did the Martians use an 11/8 interval in their music? And if they did, were there other unexpected intervals as well? We shall see below that indeed many more unexpected intervals were present in the sizes of other hexagons that can be fit to this crater. The average longitude of the west side of this hexagon is 5.9939° E (Sharonov Tower PM) which is very close to 6° E and is another example of the use of the number 6 in this crater.

Something was still missing from the hexagon fits to the 6E3S Crater and that was a hexagon that would fit the farthest extension of the large bulge that lies to the west. Fig. 10.6 shows that a hexagon with a side-to-side width of 1.6667 latitude degrees is the correct size to accomplish this. It also fit a couple of auxiliary craters. It fit the linear edge of the bottom of the western wall of a small crater lying to the east of the 6E3S Crater, and a linear section of a notch in the north side of a crater to the southeast of the 6E3S Crater. The hexagon width corresponds to the major 6th interval above a fundamental size of 1.0000 latitude degrees. The southern vertices of the east and west sides have the interesting latitude of 3.6948°°° S (Arsia Mons Prime Latitude) which has a numerical value

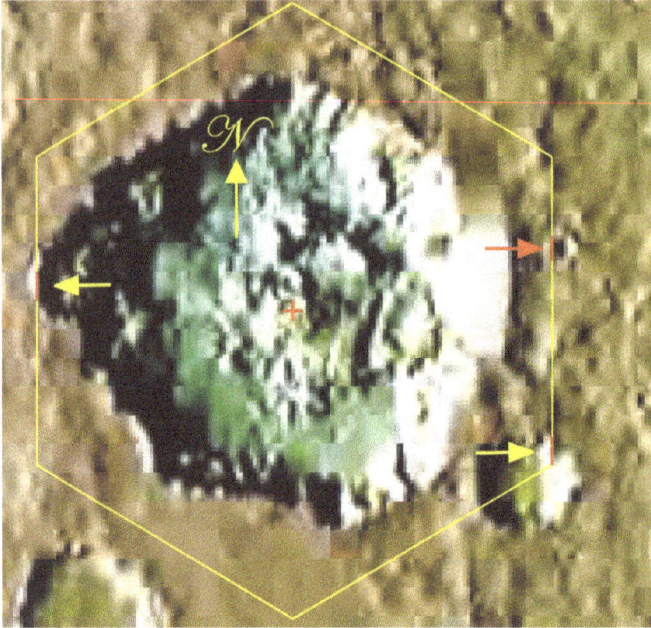

Fig 10.6: *A hexagon with a width of 1.6667 latitude degrees fits the linear perimeter of the western bulge of the 6E3S Crater. It also fits the linear edge of the bottom of the western wall of a small crater lying to the east, and a linear edge of the perimeter of a crater to the southeast of the 6E3S Crater. USGS Astrogeology.*

very close to $e^2/2 = 3.6945$. The west side has an average longitude of 5.1980°° E (Sharonov Tower PM) which has a numerical value close to $3\sqrt{3} = 5.1962$.

Light-coloured dots

On the southern perimeter and just to the north of this are 9 bright dots which may be peaks or towers (Fig. 10.7 upper). This unusual arrangement suggests some type of positional or distance referencing, but for the longest time I was unable to decipher their exact purpose. Although some of them were close to interesting coordinate values, they did not mark out meaningful coordinates in a systematic way. It also did not seem likely that so many would be used as markers of prime meridians such as the Sharonov Tower. Some of them form what may be perfect equilateral triangles (Fig. 10.7, lower figure) or perhaps isosceles triangles, but the limits of resolution prevent a definitive measurement. The larger triangle (left) has a side length of about 6.3 km, the middle triangle a side length of about 4.8 km and the smallest triangle a side length of about 3.3 km.

When I was trying to fit hexagons to linear edges in the perimeter of the 6E3S Crater with conventional music intervals, I noticed that the hexagon which had a width of 1.3333 latitude degrees (Fig. 10.3) passed through the western light-coloured dot on the southern perimeter of the

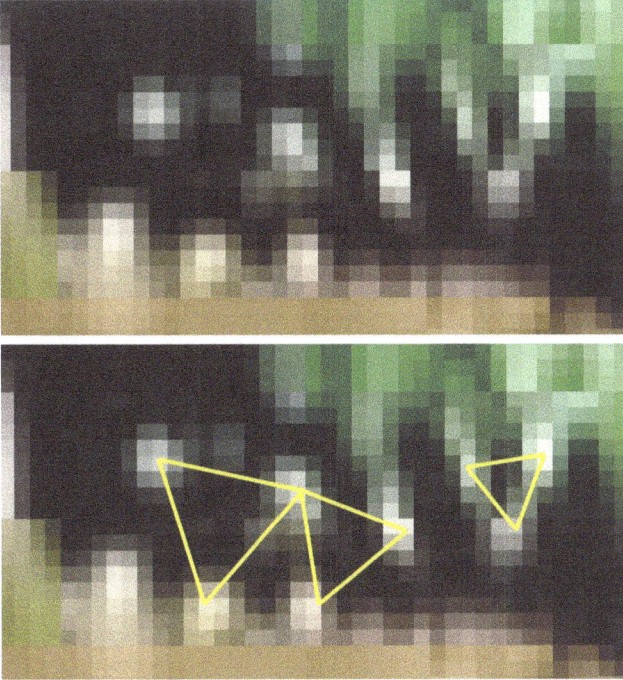

Fig 10.7: *Upper figure: nine bright spots on or near the southern perimeter suggest the presence of towers or peaks. It was very difficult to determine what purpose they might have served. Lower figure: some of the spots seem to form equilateral or isosceles triangles but the map resolution is too low to precisely determine this. USGS Astrogeology.*

crater. Also, a hexagon with a width of 1 latitude degree passed close to the centre of one of the dots lying interior to the perimeter. This stimulated my thinking. I had tried for several years now to try to figure out the purpose of these 9 dots without success. Could they perhaps serve the purpose of designating hexagons of meaningful sizes? I tried at first to fit hexagons to the interval sizes of the conventional chromatic scale but did not meet with any success. That forced me to try unconventional music intervals. But unconventional intervals are actually quite numerous, and several can give credible fits to the same structure. However, I got lucky with the middle and eastern dots on the perimeter. Hexagons that fit them well had interval sizes of 35/27 and 11/9. I chose the 35/27 interval size for the middle dot since sizes close to it had large numbers in their numerators and denominators, and I reasoned that the architects would have preferred small numbers. The interval of 11/9 was the only size that gave a reasonable fit to the eastern dot. The reason that I say I got lucky was that I noticed that both of these interval sizes had numbers related to the pentagram. The number 27 is 1/4 of 108 degrees, the internal angle of the interior pentagon (or angle between the star points) of a pentagram. The number 9 is 1/4 of 36 degrees, the angle of the star points. That inspired me to select interval sizes for the next series of dots that not only produced hexagons that gave good fits to the dots, but also contained a number related to the pentagram. Amazingly, I was

Interval Ratio	Name of Interval in Music	Width of Hexagon (Latitude Degrees)	Cents	Width of P1 Hexagon (Latitude Degrees)
125:72	augmented 6th	0.8681	955.031	0.5000
54:29		0.9310	1076.288	0.5000
108:55		0.9818	1168.233	0.5000
19:18	undevicesimal semitone	1.0556	93.603	1.0000
59:54		1.0926	153.307	1.0000
125:108	semi-augmented whole tone	1.1574	253.076	1.0000
11:9	undecimal neutral 3rd	1.2222	347.408	1.0000
35:27	septimal semi-diminished 4th	1.2963	449.275	1.0000
4:3	perfect 4th	1.3333	498.045	1.0000

Table 10.1 Hexagon sizes fitting the bright dots of the 6E3S Crater.

able to find such interval sizes for all the remaining dots! These are listed in Table 10.1 along with interval sizes in units of cents. Cents divide the octave into 1200 equal intervals and provide a convenient indicator of where you are in the octave. Note that the first 3 entries in Table 10.1 are in the octave below the reference size (P1 interval) of 1 latitude degree. The hexagon fits to the dots are shown in Fig. 10.8. Each hexagon passes through the brightest pixels of its respective dot. All of this suggests that

Fig. 10.8: *Hexagons fitted to the light-coloured dots near the southern edge of the 6E3S Crater. Each hexagon passes through the brightest pixels of the dots. The top hexagon is the smallest and has a width of 0.8681 latitude degrees. See Table 10.1 for the other hexagon sizes. USGS Astrogeology.*

the main purpose of the dots (towers/peaks) was to celebrate the pentagram.

Except for the perfect 4th, each of the interval ratios in Table 10.1 has a number in its numerator or denominator which is the size, or 1/2 or 1/4 the size, of one of the 3 angles present in the pentagram (36, 72, or 108 degrees). Both the numerator and denominator of the perfect 4th interval are factors of pentagram angles so, for example, it could be expressed as 36/27 or 72/54. The names given to most of these intervals in music are unfamiliar to all but musical experts. I could not find names for 3 of the intervals in Table 10.1 but they have all been used in one musical system or another in the past. What is interesting is that the positioning of the dots satisfies not only the requirement for an interval referring to a pentagram angle but also is an interval that is used in music. This suggests that the Martian architects were well versed in music, and used music as an essential part of their sacred geometry.

Distance Formulae to other Sites

Within the crater are 2 formations producing an "X" shape (Fig. 10.2). If the centres of the two X's are joined, and the joining line is extended outwards toward the northwest, the line eventually intersects with the AscSC2 Crater suggesting a connection with this crater. This was puzzling at first until I realized that the AscSC2 Crater and the Poynting Crater were virtually equidistant from the hexagonal centre of the 6E3S Crater, with a distance of 3650.05 km for the AscSC2 Crater vs. 3650.35 km for the midpoint of the 3 brightest pixels of the eastern arm of the arch region in the Poynting Crater (which could be used as a survey point for Ascraeus Mons - see Chapter 9, *Intelligent Mars I*). Hence, these 2 sites create an isosceles triangle with the 6E3S Crater. The closest sacred distance formula which matched the length of their distances from the 6E3S Crater was $8R'/e^2 = 3655.35$ km which was about 5 km longer (remember that R' is the northern polar radius of Mars). Most interestingly, lines drawn from the centre of the 6E3S Crater to the AscSC2 and Poynting Craters create an angle which brackets Ascraeus Mons. The angle created by a line from the 6E3S Crater to Ascraeus Mons with the line to the AscSC2 Crater is 4.57°, whereas it is 4.64° with the line to the eastern arm of the arch in the Poynting Crater. Thus a line from the 6E3S Crater to Ascraeus Mons almost perfectly bisects the angle created by the 2 lines joining the 6E3S Crater to the Poynting and AscSC2 craters (Fig. 10.9). The next step was obvious - measure the distance from the 6E3S Crater to Ascraeus Mons. It turned out to be 3199.46 km which is only 1.03 km more than the sacred distance $7R'/e^2 = 3198.43$ km. The

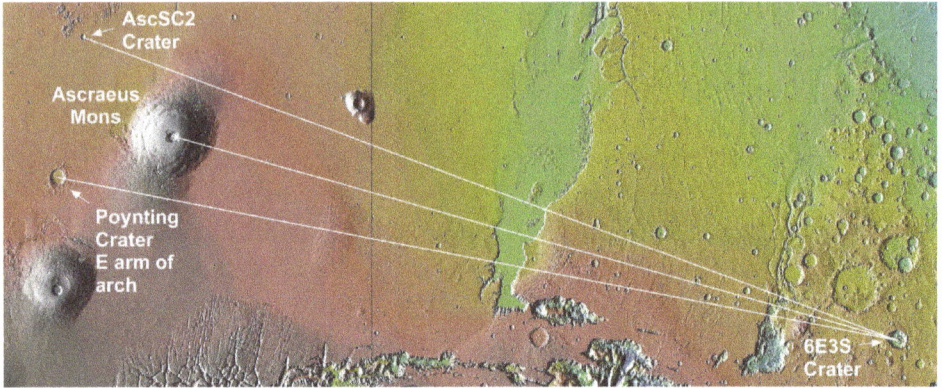

Fig. 10.9: *Ascraeus Mons sits midway in the angle created by the 6E3S Crater with the AscSC2 Crater and the midpoint of the 3 brightest pixels on the eastern arm of the arch in the Poynting Crater. USGS Astrogeology.*

symmetry between the sacred distances $8R'/e^2$ and $7R'/e^2$ piqued my interest, so I decided that I had better check out the distances from the 6E3S Crater to other major sites. Sure enough, I found 2 more sites with sacred distances having the pattern ir/e^2 where i is an integer and r is either the polar or the equatorial radius of Mars (Table 10.2). I also found 4 sites with sacred distances of the form e^2r/i, bringing the total up to 9 sites with e^2 in their sacred distance formulae from the 6E3S Crater. There is an interesting symmetry in the formulae to the AscSC2 Crater and the Poynting Crater arch with the formula to the Uranius Tholus Caldera where the $8/e^2$ term becomes inverted to $e^2/8$. Similarly, the $9/e^2$ term in the formula to the Ulysses Tholus Caldera becomes inverted to $e^2/9$ in the formula to the AscSC1a crater.

There are also sacred distances to a group of 8 sites listed in Table 10.2 which involve π^2 either in the numerator or denominator instead of e^2. Interestingly, the Nicholson and Pettit Craters share some symmetry in the use of π^2 in their sacred distance formula from the 6E3S Crater. With the Nicholson Crater, π^2 is in the numerator, and in the Pettit Crater, π^2 is in the denominator. The pentagram is referenced in 4 of the π^2 formulae with the use of the numbers 5, 9 and 18 where 5 is the number of star points in the pentagram, and 9 and 18 are 1/4 and 1/2 the size of the angle of the star points.

Finally, music ratios are used in sacred distance formulae to 4 sites. The Pavonis Mons Caldera distance (3601.54 km) from the 6E3S crater is very close to the sacred formula of $16R'/15 = 3601.28$ km. Note that 16/15 is the frequency ratio for the music interval of a minor second. This is the same interval used in the formula to the S2 crater inside the Sharonov Crater except that the interval is 1 octave lower. Also R is used instead of

Table 10.2: *Distances of sites from the 6E3S crater. R = equatorial radius and R' = northern polar radius of Mars.*

Site	Sacred	Distance (km)		
		Sacred	**Actual**	**Diff.**
Ascraeus Mons	$7R'/e^2$	3198.43	3199.46	1.03
AscSC2 Crater	$8R'/e^2$	3655.35	3650.05	-5.31
Poynting Crater E arm of arch	$8R'/e^2$	3655.35	3650.35	-5.00
Ulysses Tholus Caldera	$9R'/e^2$	4112.27	4118.00	5.73
Hecates Tholus Caldera	$20R'/e^2$	9138.38	9134.46	-3.92
Sharonov Tower	$e^2R'/14$	1781.92	1783.53	1.61
Tharsis Tholus N	$e^2R/10$	2509.46	2505.26	-4.20
AscSC1a Crater	$e^2R'/9$	2771.88	2767.65	-4.23
Uranius Tholus Caldera	$e^2R'/8$	3118.37	3121.53	3.16
Sharonov S3 Crater	$\pi^2R'/18$	1851.21	1852.92	1.71
Sharonov Triangle Tip	$\pi^2R/18$	1862.17	1859.92	-2.25
Ceraunius Tholus Caldera	$\pi^2R/11$	3047.19	3046.98	-0.21
Paros Crater	$\pi^2R/11$	3047.19	3049.87	2.68
Nicholson Crater	$\pi^2R/5$	6664.35	6657.52	-6.83
AscSC1 Crater	$9R'/\pi^2$	3078.73	3082.50	3.78
Ulysses Tholus	$12R/\pi^2$	4129.27	4127.93	-1.35
Pettit Crater	$21R/\pi^2$	7226.23	7224.73	-1.50
Sharonov S2 Crater	$(16R/15)/2$	1811.30	1811.87	0.57
Pavonis Mons Caldera	$16R'/15$	3601.28	3601.54	0.26
Alba Mons	$6R'/5$	4051.44	4057.11	5.67
Pentagram Pyramid	$4R/3$	4528.25	4530.52	2.26

R'. The distance formula to Alba Mons uses a minor third ratio (6/5) and the formula to the Pentagram Pyramid uses a perfect 4th ratio (4/3).

Summary

In summary, the 6E3S Crater has multiple themes. It brings attention to the integer divisors of 6 and 3 in its coordinates which might actually have been intended to be symmetrical by putting the latitude into a 720

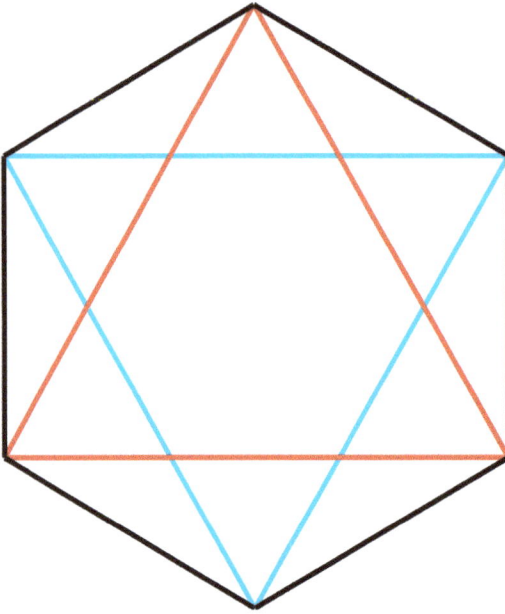

Fig. 10.10: *A pair of equilateral triangles is created by joining every vertex of a regular hexagon to the vertices which are 2 vertices away. The 2 intersecting triangles form a sacred symbol which has been used by several religions in the past and is presently used by the Jewish community to represent Judaism. The Hindu religion uses the symbol to represent the female and male aspects of the Divine with the downward pointing triangle denoting Shakti (female) and the upward pointing triangle denoting Shiva (male).*

degree system and the longitude in a 320 degree system. Then they would be 6°° E (Sharonov Triangle PM) and 6° S (720 degree system). The 6 might actually refer to the 6-sided hexagons of many sizes which fit several features of the crater such as the perimeter and the light-coloured dots. Or the 6°° E 3° S representation might refer to the 2 intersecting equilateral triangles created by joining the vertices of a regular hexagon to every vertex which is 2 vertices away (Fig. 10.10). The number 6 would represent the hexagon and the number 3 would represent the triangles. The 2 intersecting triangles have been used by several religions in the past as a sacred symbol. It is presently used by the Jewish community as the Star of David to represent Judaism. Hindus use the symbol to represent the female and male aspects of the Divine with the downward pointing triangle denoting Shakti (female) and the upward pointing triangle denoting Shiva (male). It has also been used by Satanists to represent the Beast, with the upper left and right star points denoting its horns, similar to the inverted pentagram with its 2 upper star points creating the horns of the Beast. This later use of the intersecting triangles is likely to be a perversion of a very sacred symbol which could have been used by the ancient Martians along with the pentagram to represent various aspects of the Divine. I have not come across this explicit shape on Mars either in a pyramid or a crater, but it might be implied by the hexagon-shaped craters. Another possibility is that the 6E3S crater might be referring to the 6 equilateral triangles which are obtained by joining the vertices of a hexagon to the hexagon centre. The 3 would then represent the 3 sides of each of the 6 equilateral triangles.

The light-coloured dots appear to be aligned to hexagons which are sized according to interval ratios having a numerator or denominator related to a pentagram angle. Hence, the pentagram is a strong second theme of the 6E3S Crater. This is reinforced by the fact that the hexagon centres are 54°° E of the Pavonis Mons PM. The value of 54 is one-half the size of the angle between the star points of a pentagram.

A third theme is the use of music intervals both in the size of hexagons and in 4 of the sacred distance formulae to other sites. This would imply that music is an essential component of the Martian expression of sacred geometry. The esoteric nature of some of these intervals suggests a profound depth of musical knowledge.

Another theme is the bearing angles of the straight line segments in the crater perimeter and interior which may serve a compass function due to the high number of these lines with bearing angles of 0° or ±45°. Alternatively, the bearing angles might be referring in large part to geometric figures such as the square (45°, 90°), the equilateral triangle (60°) and the pentagram (18°, 36°).

A final theme is the irrational number e which is heavily represented by the crater in the latitude position of e°°° S for the SE and SW vertices of the hexagon with a side-to-side width of 1.375 latitude degrees, and in the sacred distance formulae as e^2 for the distances from the 6E3S Crater to several important sites on Mars. The e relationship to the 6E3S Crater is reminiscent of the e relationship of the Janssen Crater and provides additional evidence that this irrational number was of critical importance to the Martian architecture. The heavy use of π^2 in distance formulae to other sites does not seem to relate to any major aspect of the 6E3S Crater, although some of the integer numbers present in these formulae represent the pentagram. Perhaps the use of π refers to its occurrence in both the pentagram and the hexagram due to the fact that their vertices lie on a circle.

There are still several unexplained notches in the perimeter and indications of intricate patterns within the crater, so there is much yet to discover about the full purpose of this very significant Martian site. Perhaps it remains unnamed so as not to draw attention to its many unnatural constructs.

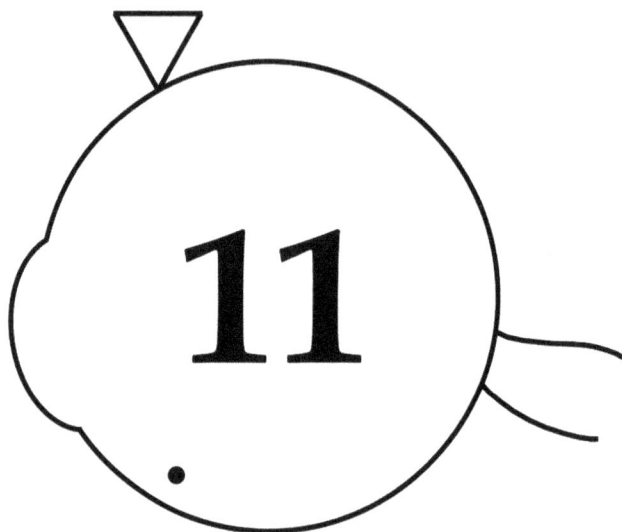

The Eye of Sharonov

O ne of the most outstanding examples of the sacred artistry and advanced engineering from amongst the many artificial Martian craters is the Sharonov Crater. This crater is located about 3500 km to the northeast of the Tharsis Montes at 301.5358° E 26.9969° N, and was named after the Russian Astronomer V.V. Sharonov (1901-1964). Note that its latitude is only 11 seconds of a degree from 27° N so it is essentially an integer number of degrees and relates to the pentagram since 27 is 1/4 the size of the angle between the star points or of the angles of the interior pentagon. We have seen this crater before in Chapter 1 as a source of the Sharonov Triangle PM and the Sharonov Tower PM. It also cropped up in Chapter 9 as interacting strongly with the Elorza Crater. This gives some indication of the huge importance of this crater, and there is much more to be revealed which I will now do in this chapter.

My first impression was that the Sharonov Crater is almost perfectly round except for a small projection outwards towards mainly the west (Fig 11.1). However, closer examination reveals that there are actually 2 main perimeters defining the outer boundaries of the crater (Fig 11.2). The outer perimeter (white circle) is what gives the crater the impression of being mostly round and its centre is what I used as the location of the coordinates for the crater itself (see above). The inner perimeter (yellow circle), while also being round, has a different centre than the outer perimeter. This causes the inner perimeter to be displaced towards the

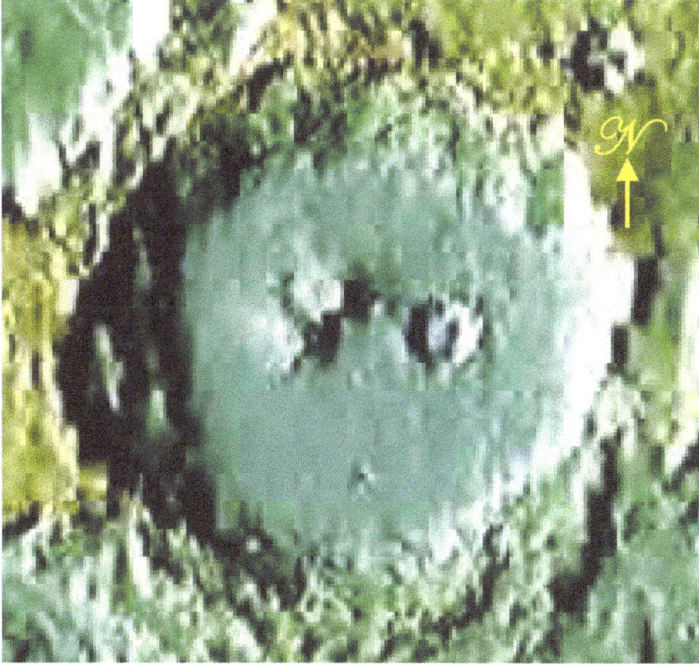

Fig. 11.1: *The Sharonov Crater is located at 301.5358° E 29.9969° N. It has a generally round appearance except for an outward projection on its western side. USGS Astrogeology.*

Fig. 11.2: *There are 2 perimeters to define the boundaries of the Sharonov Crater. The centre of the outermost perimeter (white circle) is marked by the white X placed in the centre of the crater. The inner perimeter (yellow circle) has its centre at the white spot labeled "Peak". There are also a structure (S1) and 2 small craters (S2 and S3) associated with the main crater. USGS Astrogeology.*

northeast within the outer perimeter. The centre of the inner perimeter coincides with a tiny white region in the central area of the Sharonov Crater (labeled as "Peak" in Fig. 11.2) which could be the central peak of the crater. I placed the centre of the inner perimeter in the middle of the brightest pixel for this region which should represent the highest point of the peak. It has the coordinates of 301.5905° E 27.0696° N. Its latitude is also 24.0619°° N which is close to 17√2 = 24.0416°° N. The inner perimeter is about 82 km in diameter while the outer perimeter is about 95 km in diameter so this is indeed a very large crater.

Taking a look inside, we can see a structure to the northwest of the centre of the outer perimeter. It is light-coloured which suggests that it could be a plateau rather that a crater. I have labeled it S1 in Fig. 11.2. To the east of this lies a well defined small crater about 13.1 km in diameter which is labeled S2 in Fig. 11.2. It is located just east of the centres of the 2 perimeters. A second small crater is located just outside the Sharonov Crater to its northeast. This crater is very similar in size to the S2 crater and I have labeled it as S3.

Straight Line Segments

There are many straight lines to be found in the Sharonov Crater and its smaller internal S2 crater. The southeastern perimeter of the S2 crater has a linear segment which has a bearing angle of 45° in the clockwise direction (Fig. 11.3). The presence of this line suggests that the smaller crater has been artificially constructed, at least, in part. Adding fuel to this suspicion is the fact that the small crater has a measured latitude of 27.0046° which is only 16.5 seconds more than the integer value of 27° and therefore is essentially equal to the latitude of the centre of the outer perimeter. This coordinate value is 1/4 of 108°, the size of the angles between the star points of a pentagram or of the internal angles in the embedded regular pentagon found in a pentagram. To the west of the S2 crater there are a number of structures. The first one is the "tiny" white circular or slightly oblong area (labeled "Peak") mentioned above whose centre coincides with the centre of the inner perimeter. It has a diameter of about 2 km and seems to be attached to the S2 crater with a linear dark channel which has a bearing angle of 45° in the counterclockwise direction. Moving slightly west, we observe a very large structure (bigger than the S2 crater) which is divided into 2 parts by a dark band whose northeast edge has a bearing angle of 45° in the counterclockwise direction. The northeastern part of this structure was labeled as S1 in Fig. 11.2. There is a linear edge in the eastern perimeter of the S1 structure with a bearing angle of 0°, and another linear edge in the eastern part of its southern perimeter with a bearing angle of 90°. West of the S1 structure is a linear dark band with a bearing angle of 0°. Another dark line extends westwards from this at a bearing angle of 90°. To the southwest of the S1 structure is another structure whose western boundary is very poorly demarcated and may be

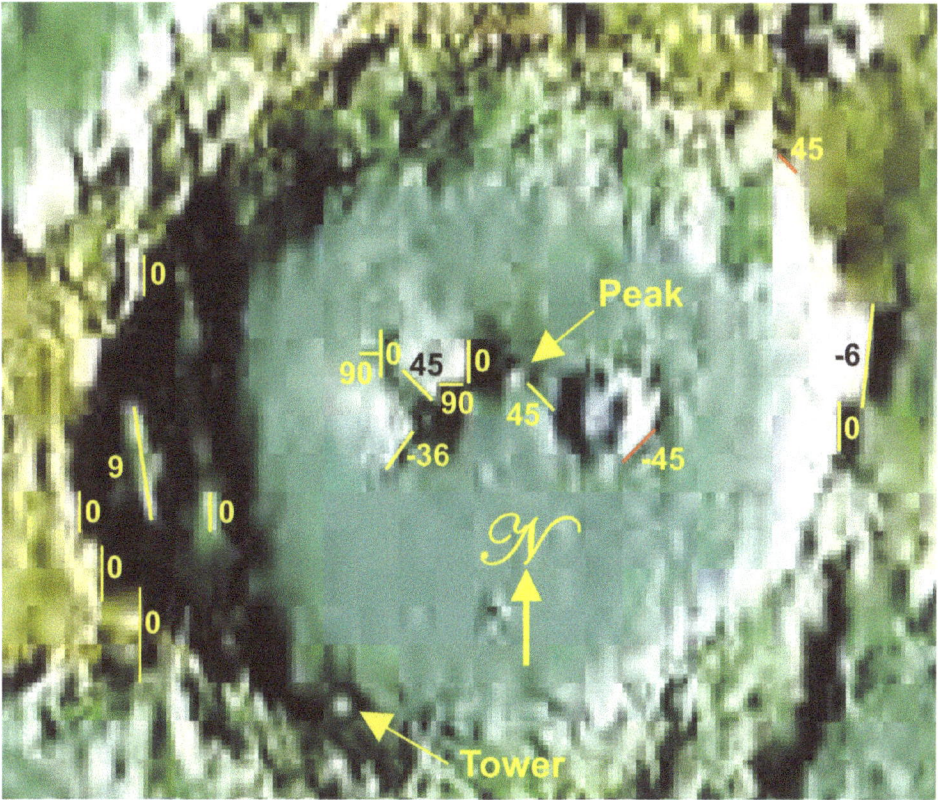

Fig. 11.3: *Many straight lines are to be found in the perimeter and interior of the Sharonov Crater. Except for north-south oriented lines, all have bearing angles which are evenly divisible by 3. A peak in the southwest corner of the crater marks the Sharonov Tower Prime Meridian. USGS Astrogeology.*

simply a sloping hill. Its southeast side, however, is quite linear and has a bearing angle of 36° in the clockwise direction.

When we examine the dark area in the far west interior of the Sharonov Crater, we see 2 straight line segments which are distinct enough to measure their bearing angles. The smaller one is aligned in the north-south direction (0°), and the longer one is tilted at a bearing angle of 9° in the counterclockwise direction. The latter also has a smaller projection to the east at its centre. Southeast of these 2 lines is another "tiny" white circular area (see arrow marked "Tower") whose coordinates are 301.1743° E, 26.3869° N. It lies between the inner and outer perimeters of the Sharonov Crater. As we have seen in Chapter 1 on prime meridians, by subtracting the longitude of the "tower" from that of Elysium Mons we arrive at a longitude of 154.0042° E which is only about 15 seconds of a degree more than exactly 154° E, putting it in integer sync with the Elysium Mons and the Dagger Midline prime meridians. Using this white circular spot as a prime meridian marker in its own right allows special numbers to be created in the longitudes of sites which could be used for sacred symbolism. The western perimeter of the Sharonov Crater has 4 linear segments, all with a bearing angle of 0°.

Going now to the eastern perimeter of the Sharonov Crater, we can see 2 straight line segments forming part of the outer edge of the crater. One points north-south (0°) and the other has a bearing angle of 6° in the clockwise direction. Just north of these 2 line segments, there is another short straight line segment of the crater perimeter with a bearing angle of 45° in the counterclockwise direction. Like the straight lines in the 6E3S Crater, the bearing angles of all of the straight lines of the Sharonov Crater, excluding the ones of 0°, are divisible by 3. The most often occurring bearing angle is 0°, making these lines potentially useful as indicators of the north-south direction.

In the region surrounding the Sharonov crater we see more evidence of artificial structures (Fig. 11.4). To the west, there are cliffs which are linear over a long distance. One has a bearing angle of 15° in the clockwise direction and another, a bearing angle of 60° in the counterclockwise direction. Northeast of the cliffs are 2 very remarkable geometric outlines

Fig. 11.4: *Lying outside the Sharonov Crater are many artificial landforms in close proximity to the crater. Their edges have long linear segments. A rectangular region and a triangular area are demarcated by trenches over parts of their borders. USGS Astrogeology.*

demarcated by trenches over parts of their borders. The first of these is a rectangle whose long axis is at a bearing angle of 45° in the counterclockwise direction. Its dimensions are about 15.4 km long by 12.2 km wide. The ratio of length to width is about 1.26 which is close to a ratio of 5/4 = 1.25, a major 3rd music interval. To the southeast of this is an equilateral triangle whose southern tip just touches the outer perimeter of the Sharonov Crater. It has a side length of about 17.7 km. The arrangement of this triangle with respect to the Sharonov Crater suggests that the triangle is a special longitude marker. However, the longitude of its tip from the Elysium Mons PM is 153.9351° E which is about 4 minutes of a degree less than an integer value. The mystery was solved in Chapter 1 with the discovery of the Dagger Peak PM which was found to be exactly 153° away from the Sharonov Triangle's tip and therefore provided validation for the prime meridian role of the triangle. To the east of the Sharonov Crater are 3 other straight line segments, 2 of which have a bearing angle of 36°, and another having a bearing angle of 6°, all in the clockwise direction. There is also a channel or groove that travels exactly in the north-south direction which is located south of the crater and has the very interesting longitude of 1.0000° E (Sharonov Triangle PM). Thus the Sharonov Crater is surrounded by a superabundance of evidence of artificiality.

The Sharonov Crater has the Shape of an Eye!

Now comes a very exciting part. After studying this crater for some time, I found that lines drawn from the Tharsis Montes to the centre of the inner perimeter had bearing angles (clockwise direction from the Montes) which were the extremely interesting integer values (within 2.5 minutes of a degree) of 70° for Ascraeus Mons, 64° for Pavonis Mons, and 60° for Arsia Mons. I was trying to decide which mountain to measure next, when it occurred to me that the western part of the Sharonov Crater which lay outside the outside perimeter had a definite curvature to it despite its rather jagged edge. Then it suddenly dawned on me. If I were to draw a curved line which followed the western edge of this area, I would create a figure which closely resembled the cross-section of an eye! The western curved line segment would create the cornea of the eye. By using a segment of an ellipse, I was able to complete the outline of the eye as shown in Fig. 11.5. I overlaid a semi-transparent diagram of the cross-section of the human eye and found that with the proper sizing it lined up very well with the new Sharonov crater outline. Then I drew in the borders of the optic nerve where it exited on the human eye diagram and found that they matched up extremely well with a lighter area of rough

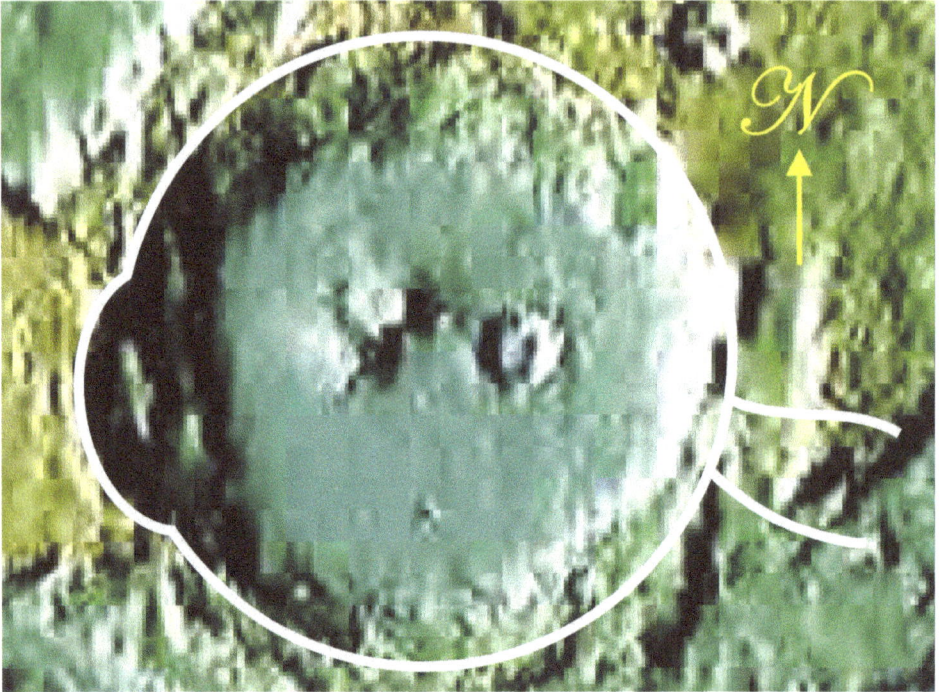

Fig. 11.5: *When the western extension of the Sharonov Crater is fitted to a curve and joined to the outer perimeter, the crater takes on the appearance of a human eye. Lines which were positioned to match the exit of the human optic nerve coincide with an area of white colored rough terrain to the east of the crater. USGS Astrogeology.*

terrain just outside the eastern perimeter of the Sharonov Crater (see white lines east of the crater). Incredibly it looks very much like we have a representation of not just an eye, but one which closely matches our very own human eye. This eye is an even better representation of the human eye than the Elorza Crater since it has a structure that corresponds to the optic nerve. Once again we have an indication that the original inhabitants of Mars might actually be our own ancestors.

This Eye Can "See"

So if this is an eye, what is it looking at? To get a good handle on this, I drew lines from the centre of the inner perimeter of the Sharonov Crater to each of the 48 sites which had been previously studied in my first book. The results are shown in Fig. 11.6 and they produce a stunning picture. It turns out that 43 of the 48 sites were "visible" through the cornea of the eye. Alba Mons, Issedon Tholus, the Issedon Tholus Caldera, the Issedon Tholus centre of the square and the Ayacucho Crater were the only exceptions. Not only were the 43 sites "visible", but this

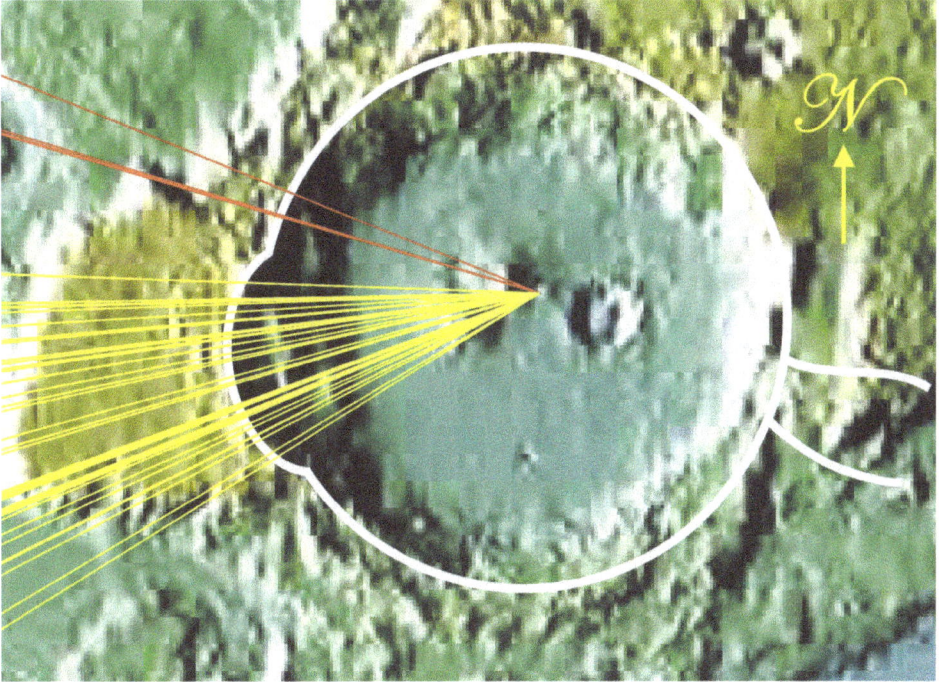

Fig. 11.6: *Field of vision of the eye of Sharonov indicated by yellow lines connecting the centre of the inner perimeter with sites on the Tharsis Rise and Elysium Rise. The red lines are connected to sites outside the view of the eye. USGS Astrogeology.*

group was centred so that the sight line of the most northerly one (Hecates Tholus Caldera) was approximately the same distance from the northern starting point of the cornea as the sight line of the most southerly one (Arsia Mons Caldera) was from the southern starting point of the cornea. The centre of the cornea surface is at 26.8331° N which is very close to the highly significant value of $12\sqrt{5} = 26.8328$.

Thus, all of the sites on both the Tharsis Rise and the Elysium Rise are under the watchful eye of Sharonov. Alba Mons, the Issedon Tholus sites and the Ayacucho Crater are to the north and therefore escape its view. This is very reminiscent of the all-seeing eye of Horus which appears on the American dollar bill and is a well-known Masonic symbol. The ultimate purpose of such an arrangement on Mars is open to speculation, but it would seem to be tied into the sacred geometry of the mountains and other sites. There are 15 out of the 48 sites which have bearing angles within 2.5 minutes of a degree of an integer or integer and one-half (see Table 11.1). The probability of such a result occurring randomly is 0.0093 or less than 1 in 100, so it is definitely statistically significant. Also shown in Table 11.1 are 2 sites other than the 48 original ones which have integer

Table 11.1: *Bearing angles (clockwise) from various sites to the inner perimeter centre of the Sharonov Crater.*

Site	Bearing (°)
From 48 Sites in *Intelligent Mars I*	
Arsia Mons	60.0414
Pavonis Mons	64.0321
Tharsis Tholus N	66.4997
Biblis Tholus Caldera N	68.5251
Ascraeus Mons	69.9796
Fesenkov Crater	78.0106
Paros Crater	82.0024
Uranius Mons	88.0322
Uranius Tholus Caldera	88.5259
Uranius Tholus Top Plateau	88.5339
Elysium Mons Caldera	89.0056
Hecates Tholus	91.9640
Hecates Tholus Caldera	92.0227
Alba Mons	105.9715
Ayacucho Crater	111.4865
Other Sites	
Dagger Peak	77.0408
Pentagon Pyramid	83.4616

or integer and one-half bearing angles to the Sharonov Crater. These are the Pentagon Pyramid and the Dagger Peak sites.

In terms of bearing angles, the eye of Sharonov seems to be concerned mostly with the Tharsis Montes and other mountains on the Tharsis Rise. Curiously, it seems that Olympus Mons is not given any emphasis here. However, its bearing angle is 82.74° so it may be that the architects were operating in a 1440 degree system instead of a 360 or 720 degree system. Other examples of such bearing angles are Jovis Tholus East (80.76°), Ulysses Tholus (68.24°), and the Nicholson Crater (74.25°). Oddly enough, the bearing angles of Alba Mons and the Ayacucho Crater sites are integer or integer and 1/2 values which suggests that the Sharonov Crater does indeed pay attention to them even though they are well out of view.

Distance Formulae to Other Sites

When we turn to the examination of sacred distances for more clues (Table 11.2), we find that the sacred distance formulae tend to be in matching groups where the formulae are similar in form and sometimes equal in value. Firstly, there are 7 sacred distances (Paros Crater, Arsia Mons, Arsia Mons Caldera, Poynting Crater, Nicholson Crater, Pettit Crater and the Pentagon Pyramid) of the form nr/i km where n is an integer, r is either the equatorial radius (R) or the northern polar radius (R') of Mars, and i is an irrational number.

Site		Distance (km)		
		Sacred	Actual	Diff.
Paros Crater	$2R/\pi$	2162.08	2161.99	-0.09
Arsia Mons	$2R'/\varphi$	4173.21	4174.51	1.30
Arsia Mons Caldera	$2R'/\varphi$	4173.21	4169.47	-3.75
Poynting Crater	$3R/\pi$	3243.12	3245.01	1.89
Nicholson Crater	$3R'/\varphi$	6259.82	6252.31	-7.51
Pettit Crater	$6R/\pi$	6486.25	6479.56	-6.68
Pentagon Pyramid	$4R/\sqrt{3}$	7843.16	7844.45	1.28
Uranius Mons	$eR'/5$	1835.49	1834.49	-1.00
AscSC1 Crater	$eR/4$	2307.95	2306.77	-1.18
Ascraeus Mons	$\varphi R'/2$	2731.40	2733.94	2.54
Ascraeus Mons Caldera	$\varphi R'/2$	2731.40	2731.63	0.23
AscSC2 Crater	eR'/π	2921.28	2919.32	-1.96
Ayacucho Crater	$8R'/15$	1800.64	1798.17	-2.47
Pentagram Pyramid	$5R'/4$	4220.25	4216.80	-3.45
Dagger Peak	$8R/3$	9056.51	9048.55	-7.95
Pavonis Mons	$\pi\varphi R/5$	3452.71	3447.94	-4.77
Issedon Tholus	$\pi\varphi R'/9$	1906.88	1905.27	-1.61
Issedon Centre of Square	$\pi\varphi R'/9$	1906.88	1906.30	-0.58
Tharsis Tholus Caldera	$\pi^2 R/17$	1971.71	1972.67	0.96
Tharsis Tholus South	$\pi^2 R/17$	1971.71	1969.35	-2.36
AscSC1a Crater	$\pi^2 R'/15$	2221.45	2219.80	-1.65
Uranius Tholus	$e^2 R'/12$	2078.91	2076.52	-2.39
Ceraunius Tholus Caldera	$e^2 R'/12$	2078.91	2077.26	-1.65
Hecates Tholus	$17R/e^2$	7813.61	7818.37	4.76
Elysium Mons Caldera	$24R/\pi^2$	8258.54	8257.67	-0.87
Albor Tholus	$24R/\pi^2$	8258.54	8262.89	4.35
Alba Mons	$2\pi R/8$	2667.36	2666.81	-0.55
Elorza Crater	$2\pi R/10$	2133.89	2131.16	-2.73
Fesenkov Crater	$\varphi\sqrt{2}R/5$	1554.26	1551.56	-2.70
Olympus Mons	$\varphi\sqrt{5}R/3$	4095.84	4088.10	-7.75
Tharsis Tholus North	$e\sqrt{3}R'/8$	1986.98	1985.57	-1.41
Biblis Tholus Caldera S	$e\sqrt{3}R/4$	3997.49	3994.82	-2.67
Poynting Crater	$eR'/(2\sqrt{2})$	3244.72	3245.01	0.29
Jovis Tholus East	$eR/(2\sqrt{2})$	3263.93	3261.95	-1.98
Uranius Tholus N Crater	$\sqrt{3}R'/(2\sqrt{2})$	2067.49	2067.81	0.32
Uranius Tholus Caldera	$\sqrt{3}R'/(2\sqrt{2})$	2067.49	2068.67	1.18

Table 11.2: *Groups of sacred distance formulae of sites from the eye of Sharonov.*

An interesting trio of sacred distances are eR'/5 km for Uranius Mons, eR/4 km for the AscSC1 crater and eR'/π km for the AscSC2 crater. Closely resembling this form is the value of φR'/2 km for Ascraeus Mons and its caldera. Then there is a set of 3 very simple distance formulae (for the Ayacucho Crater, the Pentagram Pyramid and the Dagger Peak site) which are of the form ir/j where i and j are integers, and where r = R is the equatorial radius of Mars and r = R' is the northern polar radius.

The distance of πφR/5 km from Pavonis Mons is matched by the distance of πφR'/9 km from Issedon Tholus and the centre of the Issedon Tholus Square. These are sites which are strongly related to φ, the golden mean. Note that 5 is the number of star points in a pentagram and 9 is 1/4 the size of the angle of a pentagram star point. Then we have distances from Tharsis Tholus South, the Tharsis Tholus Caldera and the AscSC1a Crater which have π^2 in the numerator and an integer in the denominator. These are matched by distances from Uranius Tholus and the Ceraunius Tholus Caldera in which e^2 is in the numerator instead of π^2. In the later pair, the distance is matched in value as well as in formula so that they form an isosceles triangle with the eye of Sharonov. The situation is flipped over for Hecates Tholus, the Elysium Mons Caldera and Albor Tholus where the integer is in the numerator and the squared irrational number is in the denominator (e^2 for Hecates Tholus and π^2 for the Elysium Mons Caldera and Albor Tholus). Note that the Elysium Mons Caldera and Albor Tholus are close to forming an isosceles triangle from the eye of Sharonov (see Table 11.2).

The distance from Alba Mons is simply 1/8th, and from the Elorza Crater, 1/10th, of the equatorial circumference. The distance of φ√2R/5 km from the Fesenkov Crater is matched by the distance of φ√5R/3 km from Olympus Mons. Next is the pair of distances of e√3R'/8 km for Tharsis Tholus North and e√3R/4 km for the southern centre of the Biblis Tholus Caldera. Finally we have a pair of distance formulae that are identical except for the fact that the northern polar radius is used for the distance from the Poynting Crater [eR'/(2√2) km], and the equatorial radius is used for the distance from the eastern centre of Jovis Tholus [eR/(2√2) km]. Similar to these formulae is the distance formula of √3R'/(2√2) km to the Uranius Tholus Caldera and a crater sitting on the northern slope of the mountain which will be discussed in Chapter 13. These 2 sites form an isosceles triangle with the Sharonov site.

In all there are 35 distances in Table 11.2 which match other distances in the form of their sacred distance formulae. Note that the Poynting Crater is assigned 2 different sacred formulae in Table 11.2 since they are very close in value and are compatible with different groupings.

Conclusion

In conclusion, both the bearing angles and distances of many of the most important sites on Mars reveal an intelligent design behind the location of the centre of the inner perimeter of the Sharonov Crater (labeled as "Peak" in Fig. 11.2). The fact that the crater can "see" most of these sites through its "cornea" bolsters the interpretation that this crater is shaped to represent an eye. The analogy of the eye breaks down somewhat with the focal point being in the middle of the eye instead of being at the anatomically correct position at the back of the eye just north of where the "optic nerve" exits. Also the focal point is at the centre of the inner perimeter rather than at the centre of the outer perimeter where the middle of the image should pass through. If the centre of the outer perimeter is used as the focal point, only 4 of the original 48 sites have a bearing angle which is with 2.5 min of an integer or integer and one-half value instead of 15 sites. If the focal point were to be moved to the back of the "eye", many more sites other than Alba Mons, the Ayacucho Crater and Issedon Tholus sites would be out of view. These latter sites were still considered important to the architects of the "eye" as seen in the integer and integer plus 1/2 values of the bearing angles of Alba Mons and the Ayacucho Crater, and in the similarity of form of some of their sacred distance formulae with other sites in view. Perhaps the Martian eye was anatomically different than the human eye. Alternatively, there is a good possibility that artistic licence was simply taken to accommodate the need to include a wider visual field. The Elorza Crater is also out of the view of the eye, and yet it is at the highly significant distance of 1/10th the equatorial circumference of Mars from the eye. It may be that the out-of-view sites were considered to be very high in status, and it would have been irreverent to be staring at them. Alternatively, it may just have been too difficult to include all of the sites in a single glance.

Although the eye of Sharonov looks westward as does the eye of Elorza, these 2 craters do not appear to be paired in the sense of forming binocular vision. The Elorza Crater is only about 45% of the size of the Sharonov Crater so they are not matched in size. They are about 2100 km from each other which is too distant to represent a binocular pair located on the front of a face. Also the visual field of the Sharonov Crater covers mainly the major mountains of Mars whereas the visual field of the Elorza Crater misses Uranius Mons, Ceraunius Tholus and Uranius Tholus but includes a region about 20 degrees south of Arsia Mons. Both eyes miss seeing Albor Tholus, Issedon Tholus and the Ayacucho Crater so they are in agreement that these sites should not be viewed. The overall purpose of the Sharonov Crater appears to be to overlook the mountain architectural sites whereas the overall purpose of the Elorza Crater appears to be to view sites in alignment with each other.

The distance of the sites in "view" of the Sharonov Crater precludes straight line sight. The curvature of the planet requires that the vision be along curved lines which, of course, violates classical physics. Hence, the viewing aspect of the crater is metaphorical only and requires the imagination to complete the picture. This is somewhat analogous to the concept of participatory sacred geometry discussed at the end of Chapter 5 where the observer is given essential starting information but has to fill in the blanks in order to complete the geometric shape or concept intended by the architect.

Like so many other sites on Mars, the Sharonov Crater is linked to the pentagram, in this case, by virtue of the latitude of the centre of its outer perimeter being at 27° N, and its 2 prime meridian sites being 54 degrees of longitude from the corresponding Pavonis Mons prime meridian sites. The number 27 is 1/4, and 54 is 1/2, of the size of the angle between the star points in a pentagram and of the size of the internal angles of the regular pentagon embedded in the pentagram's middle region. There is also a straight line segment within the crater with a bearing angle of 36°, the size of a pentagram star point angle.

The prime meridian function of the Sharonov Crater, the eye function and the pentagram reference make this one of the most important sites on Mars along with the isosceles triangle of Olympus Mons and the Tharsis giant mountains, and the north-south pointing compass formed by Albor Tholus and Hecates Tholus on the Elysium Rise. It is much more difficult, however, for the casual observer to notice its significance.

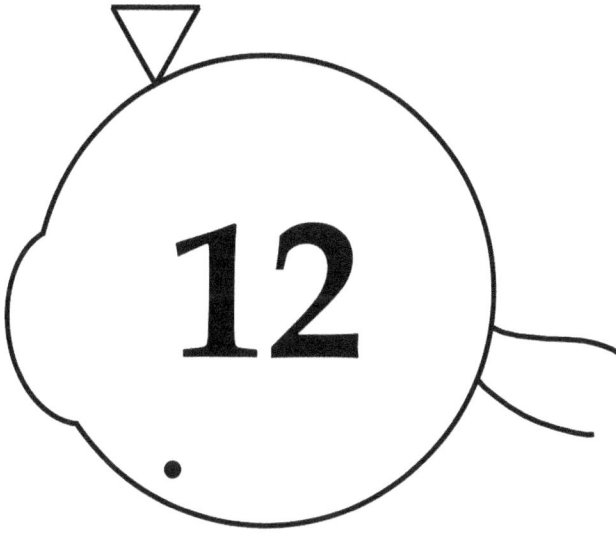

Issedon Tholus and the Ayacucho Crater

W hile the main focus of this book is on the craters of Mars, Chapter 12 will examine not only the Ayacucho Crater but also the entire mountain of Issedon Tholus together with its caldera. In *Intelligent Mars I*, Issedon Tholus was discussed as a small mountain in relation to its much more massive cousins to the south and west. It has a caldera which deserves further analysis even though it is very tiny. I also wish to take a fresh look at the Issedon Tholus site in the light of the discovery of different degree sizes and prime meridians. Just to the northeast of Issedon Tholus lies the Ayacucho Crater. This site appears to be related to Issedon Tholus and although it is called a crater by the International Astronomical Union (IAU), it may in fact be the caldera of a small mountain similar to Issedon Tholus rather than a crater.

Issedon Tholus

The coordinates of the Issedon Tholus Caldera were reported to be 265.1574° E 36.0006° N in *Intelligent Mars I*. These coordinates were of the centre of the caldera measured from a high resolution THEMIS image (Fig. 12.1). The latitude coordinate was extremely interesting since it represented a very important number in sacred geometry, namely the number of degrees in a star point of a pentagram. Since the pentagram contains several measures of the golden ratio φ, the number 36 can be

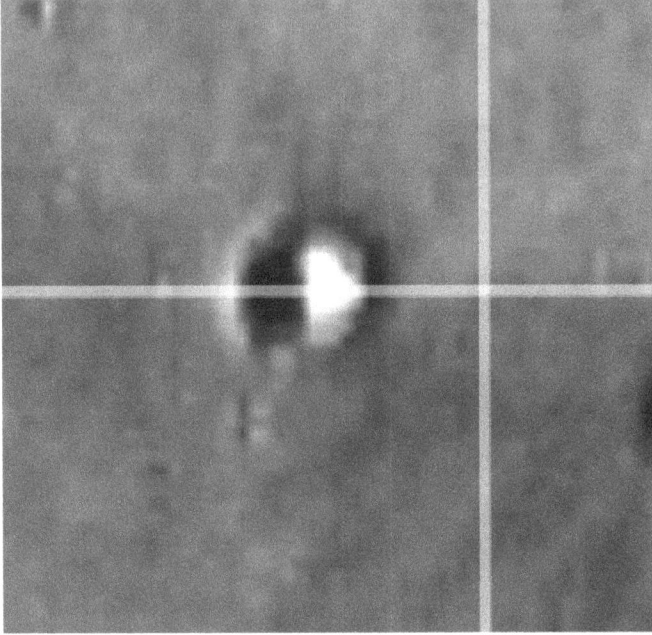

Fig. 12.1: *THEMIS image of Issedon Tholus shows the 36° N latitude line passing through the centre of the caldera. The interior of the caldera shows a white arrowhead whose tip has the remarkable coordinates of 36.0000° N 36.0007° W (Sharonov Tower PM). USGS Astrogeology.*

considered to be a proxy for φ. The longitude coordinate conveyed no meaning until I discovered the different prime meridians of Mars. When I recalculated the longitude coordinate in relation to these prime meridians, I noticed that the caldera centre was 36.0150° W of the Sharonov Tower PM. Although this differed by only 54 seconds of a degree from a pure 36°, I was puzzled as to why the architects did not bother to make it more exact.

When I first saw the THEMIS image of the crater, I was struck by how the light coloured part of the crater interior seemed to be shaped like an arrowhead. At first I simply brushed it off as an anomaly. but eventually I suddenly got the idea that just perhaps this was an intentional arrowhead and that its tip marked the position of a very important location. I determined the tip coordinates to be 265.1717 ° E 36.0000° N. When referenced to the Sharonov Tower PM, the longitude coordinate becomes 36.0007° W. Wow! Here we have a location that has both its longitude and its latitude coordinates equal to 36°. This is analogous to the coordinates of the east-west midpoint of the Pentagon Pyramid which are 12°° E of the Dagger Peak PM and 12° N. The arrowhead inside the caldera is reminiscent of the arrowheads associated with the Denning Crater (Figs. 7.2 - 7.4), especially the one protruding from the west side whose tip marks the latitude of 18° S.

The longitude coordinate of the tip of the arrow has other interesting values. It deviates less than 3 seconds of degree from 18° E and 16°° E

(PCPM), 118° E (EMPM), 117° E and 104°° E (DMPM), and 32°° W (SToPM). It is also 28.8005°°° W (SToPM) with a numerical value very close to $11\varphi^2 = 28.7984$. The longitude of 18° E from the Pavonis Caldera PM is amazing since it reinforces the relationship of this site to the pentagram (18 is half of the number of degrees of the star point of a pentagram). Together with the longitude value of 36° W (SToPM) and the latitude value of 36° N, the tip of the arrow has 3 coordinate values directly referencing the pentagram. This is a powerful indicator that the Issedon Tholus structure must have a lot to do with the golden mean φ.

If we examine this mountain closely, we can see that it is surrounded by the faint outline of what looks to be a square (Fig. 12.2). The outside edge of the mountain appears to be lying in an area of depressed terrain that changes to a higher elevation at its mostly square border. This would suggest that the base of the mountain is actually more like a square rather than like a circle, and therefore this structure must actually not be a mountain but rather a pyramid! In *Intelligent Mars I*, I fit this square with a dashed outline without revealing how I accomplished this. The readers would not have been able to understand the procedure because they had not yet read the chapter on square craters. I was only able to achieve a highly plausible fit of a square to the base of the mountain (Fig. 12.3) by drawing on my experience with square craters. I will now discuss my fitting process in full.

Except for the east vertex, I was able to align the southeast side of a

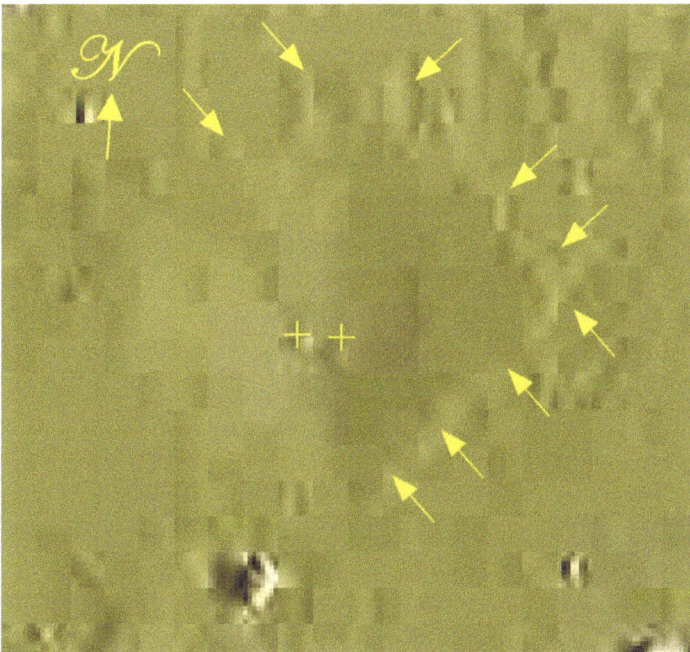

Fig. 12.2: *Issedon Tholus has a tiny caldera (left cross). Its survey centre (right cross) lies nearby. The base of this small mountain appears to be square and lies at a lower elevation than the surrounding terrain. Arrows mark parts of the faint outline of the square base. The southwest side is poorly defined. USGS Astrogeology.*

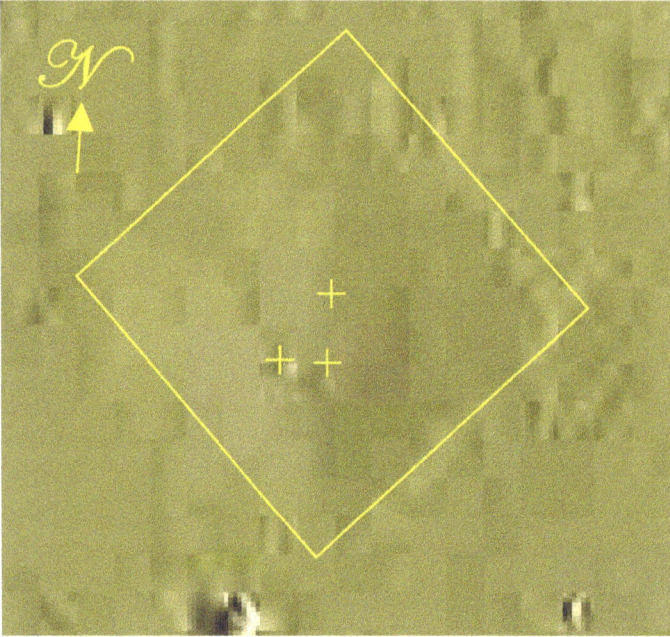

Fig. 12.3: *A square fitted to the perimeter of the Issedon Tholus. The square has a diagonal size of 1.25 latitude degrees and its northwest side has a bearing angle of 45° in the clockwise direction. The upper cross is the centre of the square. The centre of the square is Φ = 0.6180 sacred radians north of the Arsia Mons Prime Latitude. USGS Astrogeology.*

square to the edge of a stretch of terrain which is linear over a very long distance, marking an elevation change along its course. A similar situation exists for the northeast side of the Issedon perimeter except that the straight line is much fainter. The northwest side of the Issedon perimeter is even less well delineated. I adjusted the size of the square so that both its northeast and northwest sides followed the longest stretches of the change in elevation of the terrain. The border for the southwest side of the square could not be delineated so I simply drew this side based on the information present for the other 3 sides.

The border of the Issedon perimeter turned out to be consistent with a square whose diagonal size is set to 1.25 latitude degrees. According to my hypothesis that the architects used a diagonal size of 1 degree as the fundamental note in the chromatic scale of square craters, this square would be considered to be a major 3rd. Its diagonal is also 1 sacred degree in size which is in keeping with the symbolism of φ. This is because the sacred degree system has a total number of 288 sacred degrees and this can be divided into 4 sections of 72° or 8 sections of 36°. The tip of a star point in a pentagram is 36° and the 2 base angles of a star point are 72°. Hence, the use of the size of 1 sacred degree is very symbolic of φ and also of the square (288° divided into 4 parts of 72°, each representing φ, mimics 4 sides of a square). The northwest side of the square has a bearing angle of 45 degrees, the most popular orientation for square craters. Thus Issedon Tholus could be used as a compass by overhead spacecraft by fitting its outline to a square in a manner similar to what I have done.

I fine-tuned the placement of the square by giving the longitude of its centre the same longitude as the survey centre for Issedon Tholus since it was so close to that value. The latitude of the centre of the square was so close to $\Phi = 0.6180$ sacred radians north of the Asia Mons Prime Latitude that I decided that this must be its latitude. Thus I assigned the centre of the fitted square the coordinates of 265.2990° E 36.1638° N where its latitude is 35.4107°°° N (AMPL). When 35.4107 is converted to radians it equals Φ radians = 0.6180 radians, which is the inverse of φ radians = 1.6180 radians (all sacred since the latitude value is in sacred degrees). This is another indication that the Issedon Tholus square, like its caldera, is highly associated with the golden mean. Since the northwest side of the square has a bearing angle of 45 degrees, both the east and west vertices have the same latitude as the centre of the square, namely, Φ sacred radians N (AMPL). The longitude of the centre of the square is also highly connected to the golden mean being almost exactly 10φ°° E of the Pavonis Mons PM (16.1727 vs 16.1803). Although not quite as dramatic, the longitude of the centre of the square is also 105.0014°° E (Elysium Mons PM) which is only 5 seconds of a degree from being an integer number of big degrees which is evenly divisible by 5. Since the square is oriented at 45 degrees, the longitude values of both the north and south vertices of the square are also 10φ°° E (PMPM) and 105.0014°° E (EMPM). The same longitude values apply to the survey centre for Issedon Tholus.

The next most interesting coordinate of the square is the latitude value of its southern vertex which is 31.5900°° N. This number is very close to the value for $\sqrt{3}/\pi$ big radians = 31.5888 big degrees. Even more remarkable is the value of its latitude when the Arsia Mons Prime Latitude (AMPL) is used as the reference instead of the equator. Thus it is 43.6384° N (AMPL) where the numerical value is close to $acos(\varphi/\sqrt{5})$ = 43.6469. So once again there is a strong association with the golden mean.

When the latitude of the northern vertex (36.7888° N) is referenced to the Arsia Mons Prime Latitude and converted to sacred degrees it becomes 35.9107°°° N (AMPL). This number is only 1.25 minutes of a sacred degree larger than $atan(\varphi/\sqrt{5})$ = 35.8898 sacred degrees. Note the similarity to the formula for the latitude of the southern vertex. It now becomes evident that all of the vertices and centre of the square have latitude values associated with φ, the golden ratio!

Finally, the longitude of the eastern vertex of the square is 95.1748°°° E (CEPM) which has a value close to $154/\varphi$ = 95.1772. The western vertex of the Issedon square is 32.5155°° W (STrPM) which has a value close to $atan(\sqrt{3}/e)$ = 32.5047. With this, the square makes reference to all of the primary sacred numbers in its coordinates except $\sqrt{2}$, namely, φ, π, e, $\sqrt{3}$

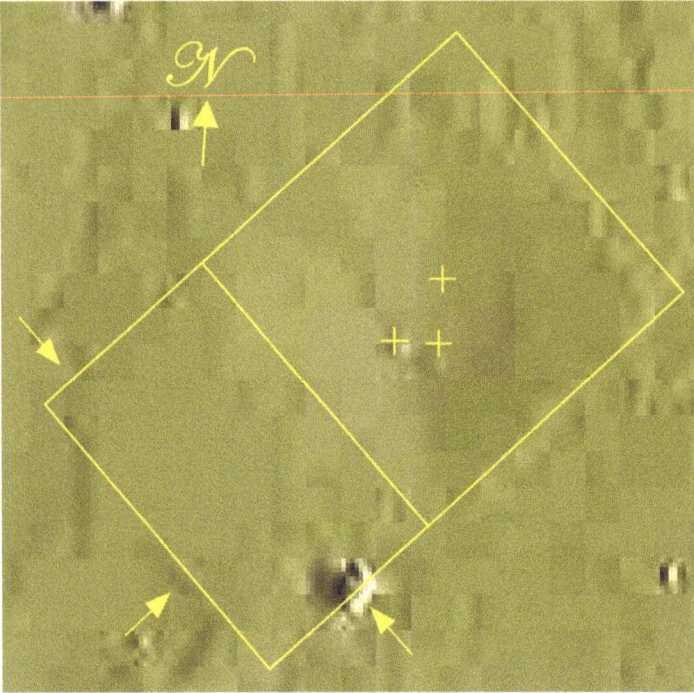

Fig. 12.4: *The square which was fit to Issedon Tholus in Fig. 12.3 can be extended to a golden rectangle. The rectangle fits the edge of an elevation change near its west vertex (upper left arrow). The southwest side fits the bottom edge of a long cliff (lower arrow left) and the southeast side runs along the bottom edge of the wall of a small crater. USGS Astrogeology.*

and √5. But since the diagonal of a square is equal to its side length multiplied by √2, then √2 is also well represented by the square.

Thus the Issedon square is a testament to all of the primary sacred numbers. However, the major emphasis seems to be on the golden mean, and there is one more extremely important finding that puts this beyond the shadow of a doubt. When we look to the southwest of the square we find that there is a sudden shift in the elevation of the terrain as was noted for the edge of the Issedon Tholus square, especially along the square's southeast and northeast borders. There is a long linear section to this change in elevation which seems to parallel the southwest border of the Issedon square. Then the thought struck me. What if I expanded the square to make a golden rectangle, a rectangle whose length is equal to φ (i.e., 1.6180) times its width? Would it fit this linear section? After making my coordinate calculations for the west and south vertices of the golden rectangle I marked them on the map and connected them to each other and to the Issedon square (Fig. 12.4). This golden rectangle aligns very nicely to the region of elevation shift in 2 locations, one near the west vertex and the other along a long linear section on the southwest side. We also see that the southeast side of the rectangle runs along the linear bottom edge of the crater wall of a small crater near the south vertex of the rectangle. Thus it looks like Issedon Tholus actually sits inside a golden rectangle rather than simply a square! If you recall, there was very little to define the southwest side of the square, so although it may be marked out, the markings are not visible at this resolution.

The coordinates of the southern vertex of the golden rectangle are 264.8266° E 35.1525° N. Its latitude is 43.2521° N and 34.6017°°° N (both AMPL) which are numbers close to $70/\varphi = 43.2624$ and $56/\varphi = 34.6099$. Its longitude is 29.0766°°° W (SToPM) and 104.6417°° E (CEPM). These numbers are close to $acos(\sqrt{2}/\varphi) = 29.0694$ and $74\sqrt{2} = 104.6518$. The coordinates of the western vertex of the golden rectangle are 264.0487° E 35.7775° N. Its latitude is extremely close to $16\sqrt{5} = 35.7771°$ N, and is 43.8771° (close to $71/\varphi = 43.8804$) and 39.0019°° north of the Arsia Mons PL. Its longitude is 103.0012°° E and 92.7010°°° E (DMPM). The latter number is close to $150/\varphi = 92.7051$. The longitude is also 15.0012°° E (PMPM) and 32.9988°° W (Sharonov Triangle PM). The latter value is close to 33, a number considered to be very important in Freemasonry.

This brings us to a discussion of the actual dimensions of Issedon Tholus. Does it extend to the borders of the golden rectangle or does it cover only a portion of the rectangle? It seems to extend beyond the southwest side of the square but it is difficult to determine if it extends right to the southwest edge of the rectangle. If we assume that it covers the entire area of the rectangle then it would have a width of 52.39 km and a length of 84.77 km. This compares to a side length of only 0.23 km for the great pyramid of Giza in Egypt. The elevation of Issedon Tholus is about 826 meters or about 5.6 times the height of the great pyramid. With these dimensions, the volume of Issedon Tholus would be about 470,000 times the size of the Giza pyramid. On the other hand, if we restrict its base size to the area of only the square part of the rectangle its volume would be about 290,000 times the size of the Giza pyramid. Either way, this is an enormous structure! I tried to resolve the extent of spread of Issedon Tholus by fitting the golden rectangle to the higher resolution THEMIS image (Fig. 12.5). It confirmed the excellent fit of the golden rectangle to the edges of the cliffs surrounding Issedon Tholus, but did not add much light to the issue of how much of this area is occupied by the tholus itself. The southwest border of the square is marked only by the edges of 3 tiny craters which are barely visible at the magnification in Fig. 12.5. The tholus seems to extend beyond this side but it is not clear as to how far it actually continues. The penetration of the tholus beyond the southwest side of the square and the lack of a clear marking for this side suggests that the square may be virtual rather than physical. Nevertheless, the square appears to be an essential component of the concept of the Issedon Tholus complex due to the special coordinate values of its centre and vertices as discussed above. The asymmetry of the position of the caldera both within

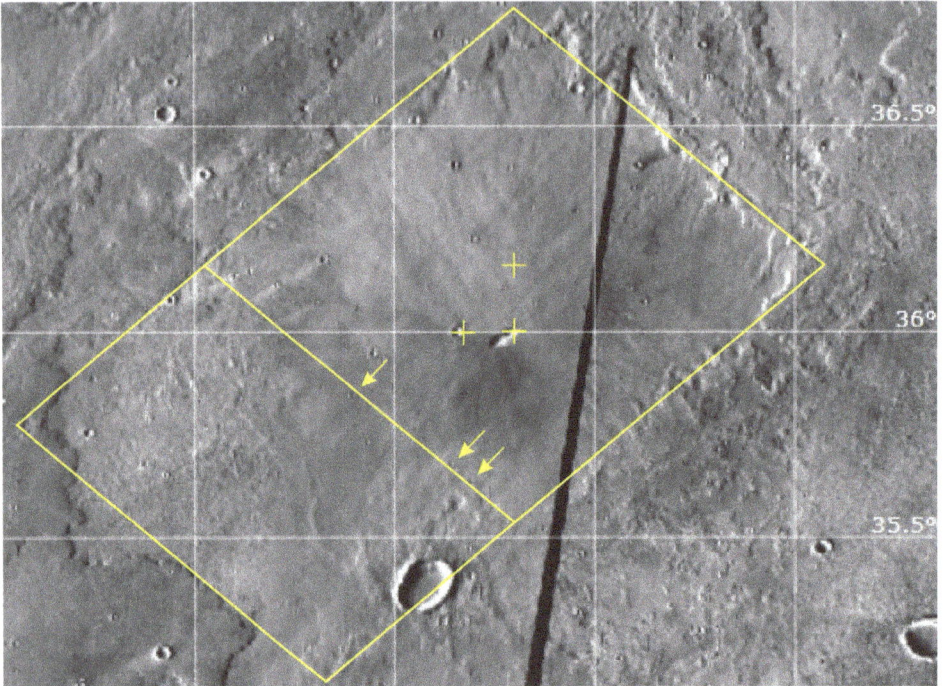

Fig. 12.5: *Fit of the golden rectangle to a THEMIS image of Issedon Tholus. Note how well the rectangle fits the boundary of the lower terrain at the edges of the Issedon structure. The square's southwest side passes along the perimeters of 3 tiny craters (see arrows) which may be visible to the reader only with a magnifying glass. USGS Astrogeology.*

the golden rectangle and especially within the square is readily apparent. The black vertical line is a region not covered by the THEMIS image.

Sacred Distances

Together with the golden rectangle enclosure, Issedon Tholus appears to be a vast monument to φ. It represents the other primary irrational numbers as well, but the overwhelming emphasis appears to be on φ. You would therefore expect this to be reflected in the sacred distances to other major sites on the planet, especially to the Pentagram Pyramid, the Pentagon Pyramid and Pavonis Mons which are highly symbolic of φ themselves. To test this out, I measured the sacred distance formulae for 45 sites studied in *Intelligent Mars I* from the Issedon sites, substituting the Issedon Caldera arrow tip for the Issedon Caldera. These are listed in Table 12.1 along with the sacred distance formulae to the midpoint of the east arm of the arch in the Poynting Crater, the crater found on the northern slope of Uranius Tholus, and the Pentagon Pyramid centre and its east-west midpoint. Each formula represents a distance which is no

Table 12.1: *Sacred distance formulae to Issedon sites relating to the pentagram.*

From Site	Distance Formula To Issedon Site (km)		
	Survey Centre	Tip of Arrow	Centre of Square
Alba Mons	$e\sqrt{5}R'/27$	$R'/(2\sqrt{5})$	
Albor Tholus	$7R'/(\sqrt{3}\sqrt{5})$		
Apollinaris Mons	eR/φ	$27R'/16$	eR/φ
Apollinaris Mons Caldera			
Arsia Mons			$\sqrt{2}\pi R/5$
Arsia Mons Caldera			
Ascraeus Mons	$\sqrt{5}R'/(3\varphi)$	$\sqrt{2}\varphi R/5$	$\sqrt{5}R/(3\varphi)$
Ascraeus Mons Caldera	$\varphi\pi R'/11$	$\varphi\pi R'/11$	$\varphi\pi R/11$
AscSC1 Crater	$\sqrt{3}\pi R/18$	$eR/9$	$\sqrt{5}\pi R/23$
AscSC1a Crater	$eR/(5\sqrt{2})$	$eR/(5\sqrt{2})$	$\sqrt{3}R/(2\sqrt{5})$
AscSC2 Crater	$\sqrt{5}\varphi R/9$	$\sqrt{3}\varphi R/7$	$\varphi R/4$
Ayacucho Crater	$eR'/(36\sqrt{2})$	$R/(8\varphi\sqrt{2})$	$\sqrt{5}R'/(27\varphi)$
Biblis Tholus			
Biblis Tholus Caldera N			
Biblis Tholus Caldera S			$e\sqrt{5}R'/8$
Ceraunius Tholus	$eR/(8\varphi)$	$eR/8\varphi$	$\sqrt{5}\pi R/33$
Ceraunius Tholus Caldera	$\sqrt{3}R'/(5\varphi)$	$\sqrt{3}R'/(5\varphi)$	$2\pi R/(18\varphi)$
Elysium Mons		$3\varphi R/e$	$3\varphi R/e$
Elysium Mons Caldera		$9R'/5$	
Fesenkov Crater	$eR/(6\varphi)$	$\varphi\pi R'/18$	$\varphi\pi R/18$
Hecates Tholus		$5R/3$	$5R/3$
Hecates Tholus Caldera	eR'/φ		
Jovis Tholus E	$e\sqrt{5}R/13$	$2\pi R'/(6\sqrt{5})$	
Jovis Tholus W	$\sqrt{5}\pi R/15$	$e\sqrt{5}R/13$	$e\sqrt{3}R/10$
Nicholson Crater		$5R'/(e\sqrt{2})$	
Olympus Mons			
Olympus Mons NE Caldera		$4R/(e\sqrt{5})$	
Olympus Mons Central Caldera	$3R/(2\sqrt{5})$		$e\sqrt{5}R'/9$
Paros Crater	$2\pi R/25$	$\sqrt{5}\pi R'/28$	$\varphi\sqrt{2}R'/9$
Pavonis Mons			$e\sqrt{5}R'/9$
Pavonis Mons Caldera			
Pentagon Pyramid Centre		$27R/16$	
Pentagon Pyramid mid EW		$27R/16$	
Pentagram Pyramid	$\sqrt{5}R/(2\sqrt{2})$		
Pettit Crater			
Poynting Crater	$\varphi\pi R'/9$	$eR/(3\varphi)$	
Poynting Crater E arm of arch	$\varphi\pi R'/9$	$eR/(3\varphi)$	
Tharsis Tholus N	$\pi^2 R'/25$	$\pi^2 R'/25$	$\sqrt{5}R/(4\sqrt{2})$
Tharsis Tholus S	$5R'/12$	$5R'/12$	$eR'/(4\varphi)$
Tharsis Tholus Caldera	$2R/5$	$e\varphi R/11$	$eR/(3\sqrt{5})$
Ulysses Tholus	$\varphi\pi R'/7$		
Ulysses Tholus Caldera		$\varphi R'/\sqrt{5}$	
Ulysses Tholus N Crater	$\sqrt{5}R/\pi$		
Ulysses Tholus SE Crater		$\varphi R'/\sqrt{5}$	
Uranius Mons	$e\varphi R/25$	$\sqrt{2}\pi R'/25$	
Uranius Tholus	$e\varphi R/25$	$2\pi R'/(22\varphi)$	
Uranius Tholus Caldera	$2\pi R/(22\varphi)$	$e\varphi R/25$	
Uranius Tholus N Caldera	$\sqrt{5}R'/(8\varphi)$	$\sqrt{5}\varphi R'/21$	$2\pi R/36$
Uranius Tholus Round Top	$\sqrt{2}\pi R'/25$	$e\varphi R/25$	$\varphi R/9$

more than plus or minus 2 km of the actual value between sites. Table 12.1 contains only those sacred distance formulae which have a component which can be considered to relate to the pentagram, and therefore, by proxy (or directly), to φ itself. Thus any formula which contained the value of φ, √5, 5, 10, 25, 9, 18, 36 or 27 was included. The value of √5 is equal to φ + 1/φ, the value of 5 is the number of star points in a pentagram, the value of 10 is the number of isosceles triangles in a pentagram, the value of 25 is the square of the number of star points in a pentagram, the values of 9 and 18 are 1/4 and 1/2 of the number of degrees (i.e., 36) in the tip of a star point of a pentagram, and 27 is 1/4 of the number of degrees (i.e., 108) between the star points of a pentagram.

Amazingly, every site except 7 in Table 12.1 has a sacred distance formula from at least one of the three Issedon Tholus sites which references the pentagram. Of the 7 sites lacking such a formula, 6 have alternative sites for the structure of interest which do have a pentagram-related distance formula. Thus Biblis Tholus has a pentagram-related distance formula for the southern centre of its caldera but not for the northern centre of its caldera or its survey centre. The calderas of Olympus Mons have formulae but not the survey site for the mountain. Arsia Mons and Apollinaris Mons have formulae for their survey centres but not for their calderas. The Pettit Crater is the only structure which does not have a pentagram-related sacred distance formula from any of the Issedon Tholus sites. Such formulae do exist for the Pettit Crater but deviate from the actual distance by more than 2 km. For example, the distance from the Issedon Caldera arrow tip is 5.6 km more than $3R/(φ√2)$ km.

The other major theme promoted by the Issedon Tholus site appears to be the square, represented by the square component of the golden rectangle. I examined this by finding sacred distance formulae to the same sites as in Table 12.1 which could be considered to represent the square. Thus any sacred distance formula containing the numbers √2, 4, 8, 16 or 32 were considered to represent the square since √2 multiplied by the side length of a square gives its diagonal value, 4 is the number of equal sides in a square, and 8, 16 and 32 are binary multiples of 4. Those formulae which deviated by no more than plus or minus 2 km from actual distances are listed in Table 12.2. Similar to the pentagram theme, most of the sites were found to be at a distance which could be formulated in terms of a sacred distance formula representing the square. The coverage was not quite as complete as with the pentagram-related equations since highly accurate formulae could not be found for 13 of the 49 sites as opposed to 7 for the pentagram. Of these, alternative sites could be found for most of the structures with the exception of Albor Tholus, Hecates Tholus, and the Pettit and Poynting craters.

Table 12.2: *Sacred distance formulae to Issedon sites relating to the square.*

From Site	Distance Formula To Issedon Site (km)		
	Survey Centre	Tip of Arrow	Centre of Square
Alba Mons	$\sqrt{2}R'/(2\pi)$	$2\pi R/(20\sqrt{2})$	$\sqrt{2}\sqrt{3}R/11$
Albor Tholus			
Apollinaris Mons		$27R'/16$	
Apollinaris Mons Caldera	$8R'/(e\sqrt{3})$		$8R'/(e\sqrt{3})$
Arsia Mons			$\sqrt{2}\pi R/5$
Arsia Mons Caldera			$4R/(\pi\sqrt{2})$
Ascraeus Mons		$\sqrt{2}\varphi R/5$	
Ascraeus Mons Caldera		$e^2R'/16$	$e^2R/16$
AscSC1 Crater		$eR'/(4\sqrt{5})$	
AscSC1a Crater	$eR/(5\sqrt{2})$	$eR/(5\sqrt{2})$	
AscSC2 Crater	$\varphi R'/4$		$\varphi R/4$
Ayacucho Crater	$eR'/(36\sqrt{2})$	$R/(8\varphi\sqrt{2})$	$\sqrt{5}R/(16e)$
Biblis Tholus	$3R/4$		
Biblis Tholus Caldera N			
Biblis Tholus Caldera S			$e\sqrt{5}R'/8$
Ceraunius Tholus	$eR/(8\varphi)$	$eR/(8\varphi)$	
Ceraunius Tholus Caldera	$eR/(9\sqrt{2})$	$e\sqrt{2}R'/18$	$eR/(4\pi)$
Elysium Mons			
Elysium Mons Caldera		$8R'/(\pi\sqrt{2})$	$8R'/(\pi\sqrt{2})$
Fesenkov Crater	$\sqrt{5}R/8$		$R/(5\sqrt{2})$
Hecates Tholus			
Hecates Tholus Caldera			
Jovis Tholus E		$4R'/(e\pi)$	$R'/(3\sqrt{2})$
Jovis Tholus W	$\sqrt{2}R'/3$		
Nicholson Crater		$5R'/(e\sqrt{2})$	
Olympus Mons			
Olympus Mons NE Caldera		$4R/(e\sqrt{5})$	
Olympus Mons Central Caldera			$3R'/(\pi\sqrt{2})$
Paros Crater	$R/4$	$R/4$	$\varphi\sqrt{2}R'/9$
Pavonis Mons			$3R'/(\pi\sqrt{2})$
Pavonis Mons Caldera			
Pentagon Pyramid Centre	$8R'/(e\sqrt{3})$	$27R/16$	$8R'/(e\sqrt{3})$
Pentagon Pyramid mid EW	$8R'/(e\sqrt{3})$	$27R/16$	
Pentagram Pyramid	$\sqrt{5}R/(2\sqrt{2})$		
Pettit Crater			
Poynting Crater			
Poynting Crater E arm of arch			
Tharsis Tholus N	$2\pi R/16$	$2\pi R/16$	$\sqrt{5}R/(4\sqrt{2})$
Tharsis Tholus S			$eR'/(4\varphi)$
Tharsis Tholus Caldera			$4R'/\pi^2$
Ulysses Tholus			
Ulysses Tholus Caldera			
Ulysses Tholus N Crater		$32R/45$	
Ulysses Tholus SE Crater			
Uranius Mons	$R/(4\sqrt{2})$	$\sqrt{2}\pi R'/25$	
Uranius Tholus	$R/(4\sqrt{2})$	$R'/(4\sqrt{2})$	
Uranius Tholus Caldera	$\sqrt{2}\pi R'/25$	$R'/(4\sqrt{2})$	
Uranius Tholus N Caldera	$\sqrt{5}R'/(8\varphi)$	$\sqrt{5}R'/(8\varphi)$	$\pi R'/(8\sqrt{5})$
Uranius Tholus Round Top	$\sqrt{2}\pi R'/25$	$\sqrt{2}R'/8$	$\varphi R'/(4\sqrt{5})$

About 65% of the formulae in Tables 12.1 and 12.2 are within ±1 km of the actual distance. To put it another way, 41 out of the 49 sites or 83.7% have at least one formula that is within ±1 km of the actual distance. This is an extraordinary degree of accuracy for so many sites, and suggests that the Issedon Tholus complex was very carefully positioned to maximize the number of sacred geometry distance formulae which represent the pentagram and the square. It also indicates that the Martian architects considered the Issedon Tholus site to be an especially important component of the planetary sculptural masterpiece.

Now let's look at the distances between Issedon Tholus sites and the Pentagram Pyramid, the Pentagon Pyramid and Pavonis Mons, all sites highly associated with φ. The Issedon Tholus survey centre is 1.60 km farther from the Pentagram Pyramid than the sacred distance of $\sqrt{5}R/(2\sqrt{2})$ = 2684.92 km. The $\sqrt{5}$ is associated with φ so that reference is obvious for both sites. The $\sqrt{2}$ would refer to the diagonal of the Issedon square. Also, not listed in Tables 12.1 and 12.2, is the sacred distance of $4R'/5$ = 2700.96 km from the Pentagram Pyramid to the centre of the Issedon Tholus square. Although it overestimates the actual distance of 2692.52 km by a rather large 8.44 km, the symbolism is powerful. The number 4 in the numerator represents the 4 sides of the square and the number 5 in the denominator represents the 5 star points of a pentagram. Moving on now to the Pentagon Pyramid, the centre of the Pentagon Pyramid is 5731.46 km from the tip of the arrow inside the Issedon Caldera, a numerical value very close to the sacred distance of $27R/16$ = 5731.07 km which represents both the pentagram and the square. The numerator integer of 27 is 1/4 the size of the angle between the star points of a pentagram and the denominator is the square of 4, the number of sides in a square. The distance from the Issedon Tholus survey centre to the Pavonis Mons Caldera is 3.52 km less than $\sqrt{5}R'/(2\varphi)$ = 2332.90 km. In this formula, the golden mean is represented by both $\sqrt{5}$ and φ. The distance of the Pavonis Mons survey centre to the centre of the Issedon square is only 0.16 km less than $e\sqrt{5}R'/9$ = 2280.16 km. The $\sqrt{5}$ represents the golden mean and 9 is 1/4 the size of the angles of the star points of a pentagram. Hence, the sacred distances from Issedon Tholus sites to the Pentagram Pyramid, the Pentagon Pyramid, and to Pavonis Mons and its caldera do indeed make reference to the pentagram and φ. The Pentagram and Pentagon pyramids also make good reference to the square.

Every one of the sites immediately south of the Issedon structure have sacred distance formulae which make reference to the Issedon square. The survey centre of Issedon Tholus is $R/(4\sqrt{2})$ km from Uranius Tholus and Uranius Mons, $eR/(8\varphi)$ km from Ceraunius Mons, and $2\pi R/16$ km from the northern peak of Tharsis Tholus. The centre of the Issedon

Tholus square is $eR'/(4\varphi)$ km from the southern peak of Tharsis Tholus and $4R'/\pi^2$ km from the Tharsis Tholus Caldera. Either the numerator or denominator of each of these sacred distance formulae is evenly divisible by 4, the number of equal sides in a square. As well, the denominator in the equal sacred distance to Uranius Tholus and Uranius Mons contains $\sqrt{2}$ which is the factor which multiplies the side length of a square to obtain its diagonal length.

There are some interesting isosceles triangles formed with Issedon Tholus sites. Thus Pavonis Mons forms an isosceles triangle with the central caldera of Olympus Mons when referred to the centre of the Issedon Tholus square. Uranius Mons makes an isosceles triangle with Uranius Tholus when referred to the Issedon Tholus survey centre. Also, the Apollinaris Mons Caldera makes an isosceles triangle with the EW midpoint of the Pentagon Pyramid when referred to the Issedon Tholus survey centre. The lengths of the 2 equal sides of the latter isosceles triangle differ by 1.25 km rather than less than 1 km for the other 2 isosceles triangles.

Not shown in Tables 12.1 and 12.2 due to space limitations are the sacred distance formulae to Issedon Tholus sites from the Sharonov Crater complex and the Elorza Crater. These are listed in Table 12.3 below and they show an especially strong pentagram link between the Issedon Tholus sites and the Sharonov Crater sites. All of the equations in Table 12.3 are within 2 km of the actual values and 11 out of 18 are within 1 km. Two equations are shown for the distance from the tip of the triangle touching the perimeter of the Sharonov Crater to the survey centre of Issedon Tholus, both within 1 km of the actual value. All of the 18 equations are related to the pentagram since they contain 1 or more of the following numbers: 5, 9, $\sqrt{5}$ and φ. Only 8 of the equations relate to the square by containing at least 1 of the numbers 4, 8, 16 and $\sqrt{2}$. However,

Table 12.3: *Sacred distance formulae to Issedon sites from Sharonov and Elorza Craters.*

From Site	Distance Formula To Issedon Site (km)		
	Survey Centre	Tip of Arrow	Centre of Square
Centre of Sharonov Crater Outer Perimeter	$\sqrt{3}\varphi R/5$	$9R/16$	$\varphi\pi R'/9$
Centre of Sharonov Crater Inner Perimeter	$\varphi\pi R'/9$	$9R/16$	$\varphi\pi R'/9$
Sharonov S2 Crater	$\varphi\pi R/9$	$4R'/\sqrt{5}\pi$	$\varphi\pi R/9$
Sharonov S3 Crater	$4R'/\sqrt{5}\pi$	$\sqrt{2}\varphi R'/4$	
Sharonov Tower	$\sqrt{3}\varphi R/5$	$9R/16$	$\sqrt{3}\varphi R/5$
Sharonov Tip of Triangle	$2R'/\varphi\sqrt{5}$		$e\varphi R/8$
	$e\varphi R/8$		
Elorza Crater	$\varphi\pi R/5$		

all of the equations to the tip of the arrow in the Issedon Tholus Caldera relate to the square as well as to the pentagram. The Elorza Crater has a sacred distance formula to the survey centre of Issedon Tholus but not to the other Issedon Tholus sites. This formula relates powerfully to the pentagram since it contains both φ and the number 5.

Sacred Bearing Angles To Issedon Sites

When the best fitting bearing angles were selected from major sites to the 3 key locations of Issedon Tholus, it turned out that the majority of good fitting bearing angles are to the Issedon Tholus Caldera arrow tip (Table 12.4). The bearing angles from the AscSC1a crater are incredible as they are simply the quantity e degrees to the Issedon Caldera arrow tip and the pure integer 3 degrees to the Issedon centre of the square with very little error, and are the same numbers associated with the torus fitting the Janssen Crater. As discussed in *Intelligent Mars I*, the Martian architects may have regarded e as an integral part of the symbolism associated with the pentagram to represent exponential growth. The number 3 is not associated with φ or the square but is related to the triangle in general (3 sides and 3 angles). These 2 numbers appear once again in the bearing angle of 3e degrees from Tharsis Tholus North to the Issedon Caldera

Table 12.4: *Sacred bearing angle formulae in the clockwise direction to Issedon sites.*

Site	Theoretical Formula	(°)	Actual (°)	Diff. (sec)
To Issedon Survey Centre				
Uranius Tholus	$10\sqrt{2}$	14.1421	14.1477	19.94
To Issedon Caldera (tip of arrow)				
AscSC1a Crater	e	2.7183	2.7277	33.82
Tharsis Tholus North*	3e	8.1548	8.1513	-12.93
Pavonis Mons	26	26.0000	26.0069	24.70
Olympus Mons Central Caldera	23e	62.5205	62.5269	22.97
To Issedon Centre of Square				
AscSC1a Crater	3	3.0000	3.0001	0.20
Apollinaris Mons	39φ	63.1033	63.1023	-3.67

*Bearing angle is in the counterclockwise direction

arrow tip. The bearing angle of $10\sqrt{2}$ degrees from Uranius Tholus to the Issedon Tholus survey centre refers to both the pentagram (contains 10 isosceles triangles) and the square (the diagonal is $\sqrt{2}$ times the side length). The bearing angle from Pavonis Mons to the Issedon Caldera arrow tip is the pure integer 26 degrees. The bearing angle from Apollinaris Mons to the Issedon centre of the square is 39φ degrees. Both of these numbers are exact multiples of 13, the perfect group size in antiquity (1 leader plus 12 followers). Also, 13 is a Fibonacci number which is related to φ since the ratio of 2 sequential Fibonacci numbers is an approximation of φ. It should be noted that the actual bearing angles listed in Table 12.4 are so close to their theoretical values that I listed their deviations in seconds of a degree in the last column.

Ayacucho Crater

The Ayacucho Crater is named after a town in Bolivia. It is called a crater, but on the THEMIS image shown in Fig. 12.6 it appears to be positioned on top of a mound which itself is situated in an area of depressed terrain.

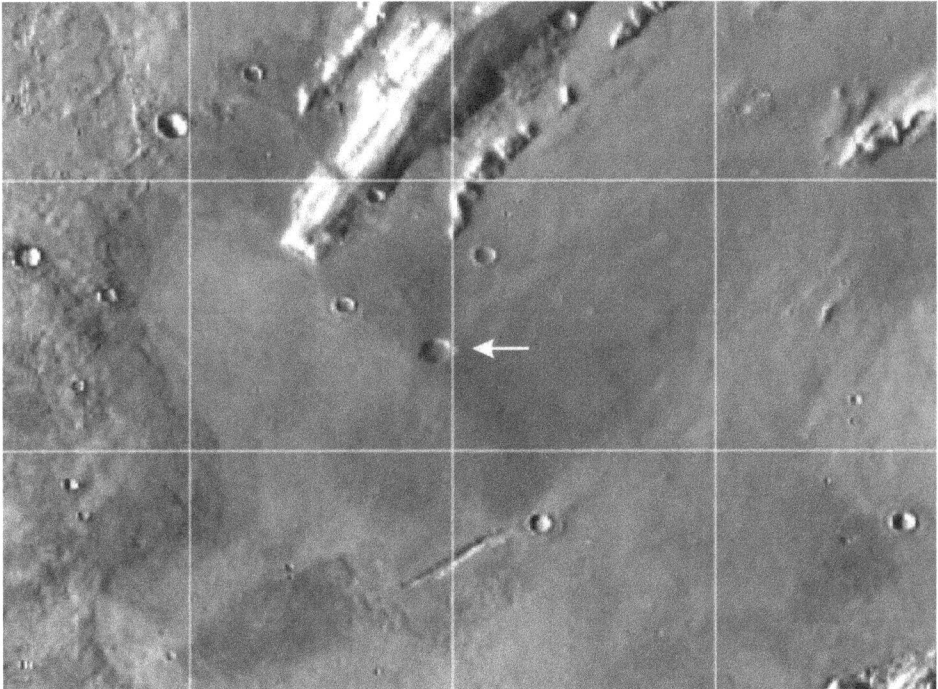

Fig. 12.6: *The Ayacucho Crater (white arrow) is located about 180 km to the northeast of Issedon Tholus. Like the Issedon Tholus Caldera, the crater appears to be situated on a small mountain or pyramid leading to the possibility that it might actually be a caldera rather than a crater. THEMIS image. USGS Astrogeology.*

This suggests that it might actually be a caldera situated on the top of a small mountain or pyramid. The fact that its perimeter does not have a rim like the other craters surrounding it also suggests that it is a caldera. The Ayacucho Crater is only about 2.5 km in diameter so it is curious that the International Astronomical Union has even bothered to give it a name.

I measured the coordinates of the Ayacucho Crater to be 267.9781° E 38.1812° N. Its latitude coordinate is very close to being equal to 38.1838° which is 27√2°. This number celebrates both the pentagram (27 is 1/4 the size of the angle between the star points) and the square (√2 multiplied by the square side length gives its diagonal length). Its latitude in terms of big degrees is 33.9388°° N which is very close to the value of 24√2 = 33.9411, once again celebrating the square. Its longitude is 106.4940°° E (DMPM), 29.5060°° W (SToPM) and 26.5016°°° W (STrPM). Although these indicate synchronization with half degrees, none of the values appear to have any special significance. However, if we probe deeper, we find that 29.5060°° W (SToPM) is very close to the value for φ/π big radians = 29.5094°°. Also with reference to the Sharonov Tower PM, the longitude in sacred degrees is 26.5554°°° W whose numerical value is close to asin(1/√5) = 26.5651. With reference to the Sharonov Triangle PM, the longitude is 29.4458°° W whose numerical value is very close to 17√3 = 29.4449. Finally, with reference to the Pavonis Caldera PM, the longitude is 18.5542°° E whose numerical value is close to asin(1/π) = 18.5607. Taking all of this together, we find that the coordinates of the Ayacucho Crater are truly remarkable in that they honour all of the basic irrational numbers except e, namely φ, π, √2, √3 and √5.

The close ties of the Ayacucho Crater to Issedon Tholus are shown by the high number of meaningful sacred distance formulae to all 3 Issedon Tholus sites. Table 12.5 lists all the meaningful formulae which are less than 1 km away from the actual distances. While it is likely that such short distances will be represented by many sacred distance formulae, normally only a few might be expected to

Table 12.5: *Sacred distance formulae from the Ayacucho Crater to Issedon Tholus sites.*

Distance Formula To Issedon Site (km)		
Survey Centre	Tip of Arrow	Centre of Square
√3R/(20φ)	√5R/(15e)	φR/(10π)
√3√5R'/72	π²R/180	R/(12φ)
eR/(36√2)	√3R'/(10π)	√5R/(16e)
√3R'/(20φ)	√2R/(16φ)	φR'/(14√5)
√3R/(12e)	R/(7φ²)	√5R'/(25√3)
eR'/(36√2)	√5R'/(15e)	φR'/(10π)
R/(6π)	π²R'/180	R'/(12φ)
	√5R'/(13π)	√5R/(27φ)
	√3πR/100	√2R'/(17φ)
	eR/50	√5R'/(16e)
	√2R'/(16φ)	√2R/(16√3)
		√3R/(21φ)

be meaningful. In Table 12.5, there are 30 such equations with 27 (90%) relating to the pentagram! They do so by the presence of the irrational numbers φ and √5, the pentagram angles of 36, 72 and 108 (as 27 which is 1/4 of 108) degrees, and the numbers 10 (the number of isosceles triangles in a pentagram), 50 (5 x 10), 100 (10 x 10) and 180 (18 x 10). However, the equations also seem to be honouring other geometric shapes as well, although not as emphatically as the pentagram. Hence, numbers relating to the square (√2, 16 = 4 x 4) are seen in 8 equations, a number relating to the equilateral triangle (√3) is seen in 9 equations and a number relating to the circle (π) is seen in 8 equations. A meaningful bearing angle formula could only be found between the tip of the arrow in the Issedon Tholus Caldera and the Ayacucho Crater. It is 74/φ = 45.7345° which is close to the measured bearing angle of 45.7406° in the clockwise direction.

The Ayacucho Crater was found to have interesting sacred distance formulae (less than 2 km away from the actual distances) to a total of 38 out of 55 sites other than the Issedon Tholus sites. Thus like Issedon Tholus, the Ayacucho Crater is well placed to create meaningful sacred distances to many sites with very small deviations from actual values. The distinctive feature of sacred distance formulae from the Ayacucho Crater to other sites was found to be in the employment of several large groups of equations which are very similar in form. Table 12.6 lists the largest of these groups together with a few smaller groups. Note that some of the sites have distance formulae which fit into more than 1 group, and that many formulae were not included due to space limitations. The largest group consists of 9 formulae having the form of eπr/n where r is either the equatorial or northern polar radius of Mars, and n is an integer. Similar to this group is a group where e is substituted by 2 and a group where e is substituted by φ. Another large group containing 7 members consists of equations of the form π²r/n where r is either the equatorial or northern polar radius of Mars, and n is an integer. The group with the simplest formulae was composed of equations of the form r/m where r is either the equatorial or northern polar radius of Mars, and m is an integer or irrational number.

The sacred distances to the 2 different peaks of Tharsis Tholus differ only in the use of the polar and equatorial radii, and in the integer value of the denominators. For Tharsis Tholus South the formula is eπR'/19 km whereas for Tharsis Tholus North it is eπR/20 km. Of the distance formulae in Table 12.6, 34 (62%) honour the circle, 28 (51%) honour the pentagram, 7 (13%) honour the square and 7 (13%) honour the equilateral triangle. Hence, although the sacred distance formulae have a strong focus on the pentagram similar to Issedon Tholus, in contrast to Issedon Tholus they focus more on the circle than on the pentagram and much

From Site	Formula	Distance to Ayacucho Crater		
		Theoretical (km)	Actual (km)	Difference (km)
Alba Mons	R/4	849.05	848.63	-0.42
Poynting Crater E arm of arch	R'/φ	2086.61	2086.32	-0.28
Olympus Mons NE Caldera	R/√2	2401.47	2402.68	1.21
Pentagon Pyramid mid EW	R'/√3	5847.75	5847.62	-0.13
Tharsis Tholus S	R/√5	1518.82	1519.29	0.47
Ceraunius Tholus	φR'/(2π)	869.43	870.50	1.07
Fesenkov Crater	eR'/9	1019.72	1020.28	0.56
Uranius Tholus Round Top	eR'/12	764.79	762.88	-1.91
Issedon Caldera arrow tip	eR/50	184.64	185.25	0.62
Issedon Tholus survey centre	eR'/51	179.95	180.83	0.88
Uranius Tholus Caldera	eR/(7√3)	761.43	761.59	0.16
Pentagon Pyramid mid EW	√3R'	5847.75	5847.62	-0.13
Uranius Mons	√3R'/(5φ)	722.82	723.23	0.41
Elysium Mons Caldera	2πR/(2√3)	6160.01	6158.23	-1.78
Jovis Tholus E	2πR'/12	1767.77	1766.31	-1.47
Paros Crater	2πR'/21	1010.16	1008.77	-1.39
Alba Mons	2πR'/25	848.53	848.63	0.10
Uranius Tholus	2πR/28	762.10	762.32	0.22
Elorza Crater	φπR/5	3452.71	3450.79	-1.92
Ascraeus Mons	φπR/10	1726.35	1726.36	0.01
Paros Crater	φπR'/17	1009.52	1008.77	-0.76
Uranius Tholus N Crater	φπR'/23	746.17	746.65	0.48
Issedon centre of square	φπR/100	172.64	174.09	1.46
Ulysses Tholus	eπR/11	2636.60	2635.66	-0.93
Olympus Mons NE Caldera	eπR'/12	2402.65	2402.68	0.02
Sharonov Crater S3 crater	eπR/16	1812.66	1812.51	-0.15
Tharsis Tholus S	eπR'/19	1517.47	1519.29	1.83
Tharsis Tholus N	eπR/20	1450.13	1450.92	0.79
Ceraunius Tholus Caldera	eπR/33	878.87	880.70	1.83
Alba Mons	eπR'/34	848.00	848.63	0.64
Uranius Tholus	eπR/38	763.23	762.32	-0.90
Uranius Mons	eπR/40	725.06	723.23	-1.83
Ceraunius Tholus Caldera	√2πR'/17	882.36	880.70	-1.66
Fesenkov Crater	√3πR'/18	1020.62	1020.28	-0.35
Issedon Caldera arrow tip	√3πR/100	184.80	185.25	0.45
Ulysses Tholus	√5πR'/9	2635.24	2635.66	0.42
Alba Mons	√5πR'/28	847.04	848.63	1.59
Ulysses Tholus N Crater	√5πR/32	745.55	746.65	1.10
Uranius Mons	√5πR/33	722.96	723.23	0.27
AscSC1a Crater	π²R'/23	1448.77	1447.97	-0.81
Paros Crater	π²R'/33	1009.75	1008.77	-0.98
Ceraunius Tholus Caldera	π²R/38	882.08	880.70	-1.38
Uranius Tholus Caldera	π²R/44	761.80	761.59	-0.21
Uranius Tholus N Caldera	π²R/45	744.87	746.65	1.78
Uranius Mons	π²R'/46	724.39	723.23	-1.15
Issedon Caldera arrow tip	π²R'/180	185.12	185.25	0.13
Jovis Tholus E	φ²R'/5	1767.80	1766.31	-1.49
Tharsis Tholus Caldera	φ²R'/6	1473.17	1472.78	-0.38
AscSC2 Crater	√2φR'/5	1545.12	1546.76	1.65
Elysium Mons	√5φR/2	6143.77	6144.70	0.94
Sharonov Crater tip of arrow	√5φR/7	1755.36	1757.29	1.93
Uranius Tholus Round Top	√5φR'/16	763.45	762.88	-0.57
Uranius Mons	√5φR/17	722.80	723.23	0.44
Tharsis Tholus N	√2eR/9	1450.64	1450.92	0.28
AscSC1 Crater	√2√5R/9	1193.30	1192.46	-0.84

Table 12.6: *Sacred distance formulae from the Ayacucho Crater to planetary sites.*

Table 12.7: *Bearing angle formulae to the Ayacucho Crater.*

| | Bearing angle to Ayacucho Crater | | | |
| | Theoretical | | Actual | Difference |
Site	Formula	(°)	(°)	(°)
Paros Crater	18	18.0000	17.9920	-0.0080
AscSC1 Crater	20	20.0000	20.0055	0.0054
Pavonis Mons Caldera	27	27.0000	26.9878	-0.0122
Jovis Tholus East	48	48.0000	47.9793	-0.0207
Tharsis Tholus North	$e/\sqrt{3}$	1.5694	1.5487	-0.0207
Tharis Tholus Caldera	$4/\varphi$	2.4721	2.4633	-0.0089
AscSC2 Crater	$\operatorname{asin}(\sqrt{2}/\sqrt{5})$	39.2315	39.2323	0.0008
Biblis Caldera N centre	$\operatorname{asin}(\sqrt{2}/\sqrt{5})$	39.2315	39.2055	-0.0260
Biblis Tholus	$\operatorname{atan}(\sqrt{5}/e)$	39.4408	39.4351	-0.0057
Olympus NE caldera	$\sqrt{3}/\varphi$ radians	61.3332	61.3596	0.0264
Sharonov Crater arrow tip	$40\sqrt{3}$	69.2820	69.2880	0.0060
Elysium Mons Caldera	51φ	82.5197	82.5174	-0.0023

more on the circle than on the square.

Remarkably, there are 4 bearing angles from various sites to the Ayacucho Crater which are close to pure integers (Table 12.7). The integer value of 18 from the Paros Crater is 1/2 the size of the angle of a pentagram star point. The integer value of 27 from the Pavonis Mons Caldera is 1/4 the size of the angle between the star points of a pentagram or of the interior angle of a regular pentagon. The integer 27 is also equal to 3^3 which is a huge reference to the number 3, the number of sides in a triangle. The integer of 20 from the AscSC1 Crater and the integer of 48 from Jovis Tholus East are evenly divisible by 4 which could be a reference to the square. The value of 20 is also evenly divisible by 5 which could be a reference to the pentagram.

There are 8 other bearing angles which are very close to sacred formula values. In 6 of these, the use of φ and $\sqrt{5}$ suggest a reference to the golden mean. In 4 of these also, the use of $\sqrt{2}$ and the number 4 (including $40 = 4 \times 10$) suggests a reference to the square. The use of $\sqrt{3}$ in 3 of the formulae suggests a reference to the equilateral triangle. So once again we see an emphasis on the pentagram. However, there is no reference to the circle, but rather a reference to the square and equilateral triangle.

Summary

It would seem that the Issedon Tholus and the Ayacucho Crater work as a team to represent primarily the pentagram. Both of these sites make heavy reference to this geometric figure by emphasizing pentagram-related irrational numbers and integers in coordinates, sacred distance formulae and bearing angles. The pentagram is represented especially by the irrational numbers φ and √5, by the integers 108, 72 and 36 (and their binary divisions: 54, 27, 18 and 9) which are the sizes of angles found in the pentagram, and by the integers 5 (number of star points), 10 (the number of isosceles triangles in a pentagram), 25 (5 x 5), 50 (5 x 10), 100 (10 x 10) and 180 (18 x 10). The pentagram might also be represented by e since the pentagram is a symbol of exponential growth (see *Intelligent Mars I*, Chapter 9). The supreme indication that Issedon Tholus represents the pentagram comes from the values of the coordinates of the tip of the arrow found in its caldera: 18° E (PCPM), 36° W (SToPM) and 36° N. The number 36 is the number of degrees in the star points of the pentagram and 18 is one-half this value. Issedon Tholus is also enclosed in a golden rectangle whose length/width ratio is equal to φ which is a strong reference to the pentagram. The cooperation of the Ayacucho Crater in representing the pentagram in tandem with Issedon Tholus is seen in the large number of close fitting sacred distance equations between these 2 structures in which 90% of them refer to the pentagram (Table 12.5). With reference to other major sites, the Ayacucho Crater also promotes the pentagram in sacred distance formulae and in some of the integer bearing angles and bearing angle formulae.

Issedon Tholus emphasizes the square in sacred distance formulae with the numbers √2, 4, 8 and 16 especially to sites south of the Issedon structure. Since Issedon Tholus sits in a golden rectangle enclosure of which the square is an important component, the emphasis on the square is understandable. This emphasis is not shared by the Ayacucho Crater which emphasizes the circle in sacred distance formulae containing the value π. The circle could refer to the equilateral triangle, the square and the pentagram since the outer vertices of all 3 of these geometric shapes lie on a circle. The Ayacucho Crater is also notable in its use of equation categories which are composed of equations very similar in form such as $e\pi r/n$ where r is either the equatorial or northern polar radius of Mars, and n is an integer.

The precision of the coordinates and the very low errors in the sacred distances to a very large number of major sites on Mars (Tables 12.1 - 12.3 and 12.5 – 12.6) suggests that Issedon Tholus and the Ayacucho Crater are two of the most important sites in the Martian architecture devoted to the manifestation of sacred geometry despite their small size relative to the giant mountains.

13

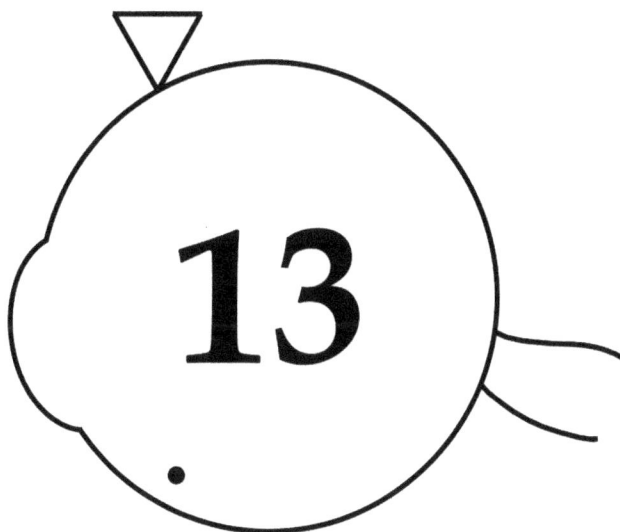

Craters on the Mountains

Several craters of substantial magnitude can be found on the giant mountains of Mars. I suspected for quite some time that at least some of them were artificial but I was too busy investigating other matters to systematically investigate them. Early on, however, I did try many times to unlock the secrets of the 2 large craters on Ulysses Tholus as they looked very much to me to have been artificially created. My efforts did not meet with any success so I eventually abandoned my study.

With the discovery of the Martian prime meridians and the existence of several systems used to measure out degrees on Mars, I was now equipped to undertake a proper investigation of these craters. I also had built up much experience about the ways in which the Martian architects expressed their approach to sacred geometry. I restricted myself to the most obvious and largest craters since I figured that these would have the greatest probability of being created for spiritual or navigational purposes if they were indeed artificially constructed. Since it was the craters on Ulysses Tholus that piqued my interest first and presented a puzzle that I was unable to solve for the longest time, I will start there. Those unfamiliar with the layout of the mountains which are discussed in this chapter can refer to the figure at the back of the book to orient themselves.

Fig. 13.1: *Ulysses Tholus has 2 large craters on its edifice. The ratio of the area enclosed by the perimeter of the entire mountain to the area of its huge caldera is approximately equal to π. The centre of the northern crater is located very close to a latitude of π°° N and a longitude of 50°°° W of the Sharonov Triangle PM. USGS Astrogeology.*

Ulysses Tholus

Ulysses Tholus (Fig. 13.1) was discussed in *Intelligent Mars I* mostly in terms of its relation to the Pentagram Pyramid, and it was pointed out that the perimeter of the mountain is largely composed of straight line segments rather than being truly round. Ulysses Tholus is curious in that the total area enclosed by the perimeter of the mountain (approximately 7605 km^2) is very close to being π times the area of its huge caldera (approximately 2476 km^2). The ratio of the 2 areas by my measurements is 3.0715 rather than 3.1416. When you add in the areas of its 2 large craters, you find that almost 42 % of the mountain as viewed from above is either caldera or crater. The artificiality of this mountain is further suspected when you examine the coordinates of the centre of its northern crater which were reported to be 238.5874° E 3.5284° N in *Intelligent Mars I*. It turns out that these coordinates are very close to 50°°° W (Sharonov Triangle PM) and π°° N. I decided to examine what would happen if I tried to fit a circle to the crater perimeter using these exact values for the circle centre. The left picture in Fig. 13.2 shows the fit of the crater to a circle large enough to reach both east and west sides of the crater. This was the circle used to determine the coordinates listed in *Intelligent Mars I*. The large size of the circle caused it to exceed the crater perimeter in some regions, particularly on the southern side. By fitting a different

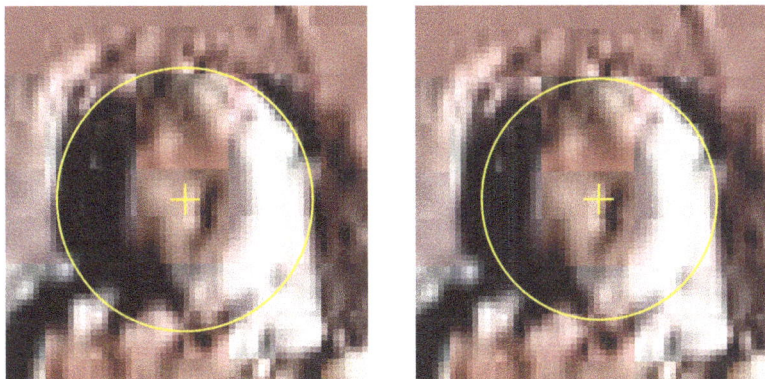

Fig. 13.2: *Left: Northern crater on Ulysses Mons fit to a circle that touches the perimeter on both the west and east sides. Right: The northern crater on Ulysses Tholus fit to a circle whose centre is set to 50°°° W (Sharonov Triangle PM) and π°° N. This circle shows that the crater is asymmetrical on its west side. USGS Astrogeology.*

circle to the perimeter in which the centre of the circle was fixed to the exact coordinates of 50°°° W (Sharonov Triangle PM) and π°° N, the result in the right picture of Fig. 13.2 was achieved. With this fit, the northern and southern portions of the crater perimeter meet the circle properly and the asymmetrical nature of the crater is revealed. The centre of the circle now is located on top of what appears to be the central peak of the crater. Its location gives Ulysses Tholus a second important marker of π in addition to the ratio of the mountain area to the caldera area.

The coordinates (238.6047° E 3.5343° N) of the new centre of the northern crater on Ulysses Tholus also encode $\sqrt{2}$, e and φ. Its latitude of 3.5343° N is very close to $5/\sqrt{2} = 3.5355$, and is 2.8274°°° N which is very close to $2\sqrt{2} = 2.8284$. Its longitude is 81.2732°° E (Elysium Mons PM) which is close to $11e^2 = 81.2796$, and is 55.6157°° W (Sharonov Tower PM) which is close to $90/\varphi = 55.6231$. Its longitude is also 6.8000°°° W (Pavonis Mons PM) which is very close to $11/\varphi = 6.7984$.

The second crater on Ulysses Tholus lies on the mountain's southeast side, and like the northern crater, covers almost the entire width of the side between the mountain and caldera perimeters. The coordinates of this crater (239.0599° E 2.6432° N) show a connection to $\sqrt{2}$ and π. In terms of big degrees, the latitude is 2.1146°° N which is close to $3/\sqrt{2} = 2.1213$ and the longitude is 81.6778°° E (Elysium Mons PM) which is close to $26\pi = 81.6814$. The longitude is also 55.1509°° W (Sharonov Triangle PM) which is close to $39\sqrt{2} = 55.1543$. Note that both 26 and 39 are divisible by 13, a Fibonacci number and the perfect group size in antiquity. This crater takes on great importance in its relationship to the northern crater (Fig. 13.3). The distance between the 2 craters (59.29 km) is almost exactly equal to 1

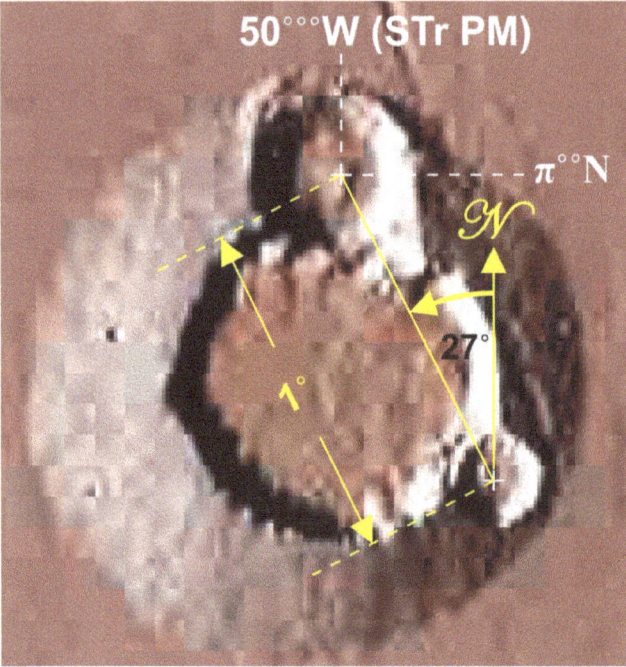

Fig. 13.3: *The centre of the northern crater on Ulysses Tholus is 50°°° W of the Sharonov Triangle PM and π°° N of the equator. The imaginary line connecting the northern crater centre to the southeast crater centre has a counter-clockwise bearing angle of 27° and is 1 latitude degree in length. USGS Astro-geology.*

latitude degree or 59.27 km. The bearing angle from the SE crater to the northern crater is equal to 27.0253° in the counterclockwise direction which is close to 27°. The value of 27 is very significant since it not only emphasizes the number 3 as it is equal to 3^3 but also is 1/4 of the size of the angle between the star points of a pentagram.

Thus the 2 craters on Ulysses Tholus have important sacred geometry characteristics. Not so obvious is the possible use of these landmarks for navigational purposes. The length of the distance between craters could be used by a spacecraft to determine its altitude as well as to calibrate for distance on the planetary surface. The bearing angle between them could provide directional information. Directional information can also be achieved by lining up the centres of the caldera (238.6052° E) and the northern crater (238.6047° E) to obtain a pointer to the north and south poles since their longitudes are virtually identical.

Olympus Mons

There are 3 craters on Olympus Mons which stand out (Fig. 13.4). The Pangboche Crater (226.6172° E 17.2641° N) and the Karzok Crater (228.2655° E 18.3945° N) lie close to the caldera complex, and the 3rd crater (228.6052° E 21.9004° N) is found just south of the northeast edge of the mountain edifice. Meaningful coordinate values were found for all 3 craters. The latitude of 17.2641° N for the Pangboche Crater is close to

Fig. 13.4: *The Karzok Crater is located east, and Pangboche Crater south, of the Olympus Mons Caldera. A third unnamed crater is located close to the northern edge of the mountain.   USGS Astrogeology.*

$10\sqrt{3} = 17.3205°$ N. Its longitude (70.6778°° E, Crater Edge PM) is quite close to $26e = 70.6753$ and $27\varphi^2 = 70.6869$. Note that 26 is twice 13 which is the traditional group size for governmental perfection, and 27 is 1/4 the size of the angle between the star points of a pentagram. The latitude of 18.3945° N for the Karzok Crater is close to $13\sqrt{2} = 18.3848°$ N giving us the number 13 once again. With reference to the Arsia Mons Prime Latitude, the latitude of the Karzok Crater is 23.5503°° N which is close to $9\varphi^2 = 23.5623$. Note that 9 is 1/4 the size of the angle of a pentagram star point. The longitude of the Karzok Crater is 71.1939°° E (Dagger Midline PM) which is very close to $44\varphi = 71.1935$. It is also 18.8392° W (Pavonis Mons PM) which is close to $6\pi = 18.8496$, and 81.1608° E (Crater Edge PM) which is very close to $31\varphi^2 = 81.1591$.

There are interesting coordinates for the northern crater as well. While there is nothing remarkable in its latitude from the equator, when measured from the Arsia Mons Prime Latitude, it turns out to be 30.0000° N or 24.0000°°° N. Its longitude is 80.5005° E and 71.5560°° E of the Dagger Peak PM. These numbers are very close to $36\sqrt{5} = 80.4985$ and $32\sqrt{5} = 71.5542$ where 36 is the size of the angle of a star point of a pentagram and 32 is 2^5. When rounded to 3 decimal places, the longitude of the northern crater is also 72.500° W or 58.000°°° W of the Sharonov Triangle PM.

The 3 craters show important relationships to each other (Fig. 13.5). The distance of 208.67 km from the Karzok Crater to the northern crater almost exactly equals $R'/(10\varphi) = 208.66$ km where R' is the northern polar radius

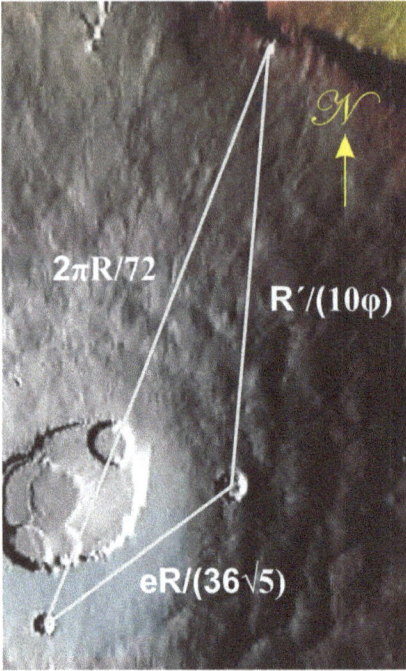

Fig. 13.5: *The distance formulae between Olympus Mons craters emphasize the pentagram and the golden ratio. USGS Astrogeology.*

of Mars. The distance of 296.38 km from the Pangboche Crater to the northern crater is virtually identical to $2\pi R/72$ = 296.37 km. Finally the distance of 114.63 km between the Karzok and Pangboche craters is very close to $eR/(36\sqrt{5})$ = 114.68 km. These formulae focus heavily on the pentagram with the numbers φ, 10, 36 and 72. The number 36 is the size of the angle of a pentagram star point and 72 is the size of the base angles of a pentagram star point. Note also the remarkable occurrence of $36\sqrt{5}$ in both the longitude (DPPM) of the northern crater and the distance formula between the Karzok and Pangboche craters.

The clockwise bearing angle of 5.1966° between the Karzok Crater and the northern crater is almost identical to $3\sqrt{3}$ = 5.1962°. The clockwise bearing angle of 21.9914° between the Pangboche Crater and the northern crater is virtually identical to 7π = 21.9911 and is also very close to the integer value of 22. The clockwise bearing angle of 54.2306° between the Pangboche and Karzok craters is about 7 minutes of a degree more than the value of $\text{atan}(\sqrt{5}/\varphi)$ = 54.1102. The formulae for the bearing angles together with the formulae for the distances between craters make reference to all the primary sacred numbers except for $\sqrt{2}$, and that number is found in the latitude of the Karzok Crater as mentioned above. The greatest emphasis of these 3 craters, however, seems to be on the pentagram.

Ascraeus Mons

There are 2 craters on the eastern side of Ascraeus Mons near the outer edge of the mountain (Fig. 13.6). The centre of the more southern crater (257.6734° E 9.4004° N) was found to be 17.5000° or 14.0000°°° north of the Arsia Mons Prime Latitude. Its latitude is also 15.5556°° N (AMPL) which is very close to $11\sqrt{2}$ = 15.5563. Its longitude is 110.5010° E (Elysium Mons PM), 109.5010° E (Dagger Midline PM), 43.4990° W (Sharonov Tower PM) and 10.5010° E (Pavonis Caldera PM). Besides being very close to integer and 1/2 numbers, the latter 2 values are almost equal to 16e = 43.4925 and 17/φ = 10.5066. The crater's longitude is also 109.5687° E (Dagger Peak PM) which is very close to 49√5 = 109.5673.

The eastern crater (258.7142° E 12.3180° N) has a latitude of 9.8544°°° N which is close to π^2 = 9.8696°°° N. It is 33.9666°°° W of the Sharonov Tower PM which is close to 21φ = 33.9787. The distance between the craters is 183.25 km which is close to eR'/50 = 183.55 km, and to

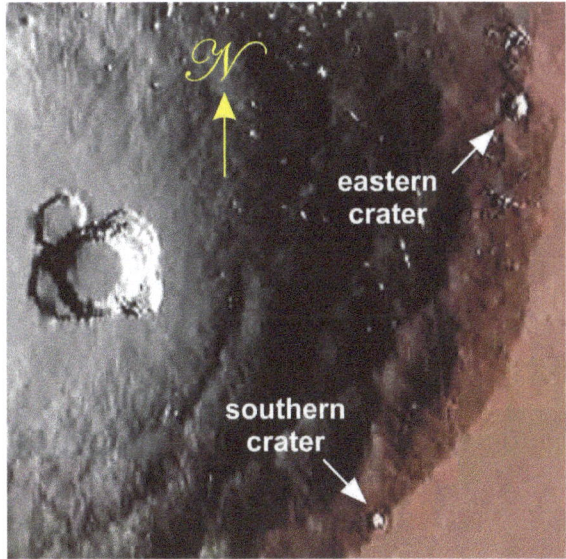

Fig. 13.6: *Craters on the eastern side of Ascraeus Mons. USGS Astrogeology.*

φR/30 km or πR/(36φ) km, both of which = 183.17 km. The bearing angle between the southern and eastern craters is 19.3055° which is about 6.65 minutes of a degree less than the value for 12φ = 19.4164°. This value of the bearing angle shows incredible symmetry with the latitude of Ascraeus Mons which is 19.3787° N (AMPL), a numerical value which is also close to 12φ. Overall, while these 2 craters on Ascraeus Mons honour e, √2 and π, they seem to emphasize the pentagram both in the formulae for the coordinate values of the craters (contain φ and √5) and in the formulae for distance and bearing angle of the imaginary line that joins the 2 craters (contain φ, 36 and 50 = 5 x 10).

Uranius Tholus

Neither Pavonis Mons nor Arsia Mons have significant craters on their surfaces so the next mountain I am going to consider is Uranius Tholus. There is a crater on the north side of this mountain and another one on the east side (Fig. 13.7). There is also a large crater abutting the western edge of the mountain which I decided to investigate due to its size even though it is not on the mountain. The coordinates of the eastern crater are 262.9243° E 26.2683° N. Its latitude expressed in sacred degrees is 21.0146°°° N which is close to 21 and to 34/φ = 21.0132. Note that 21 is a Fibonacci number. It also has a latitude of 27.4943°°° N (AMPL) which has a numerical value close to 27.5. Its longitude is 33.9983°° W (Sharonov

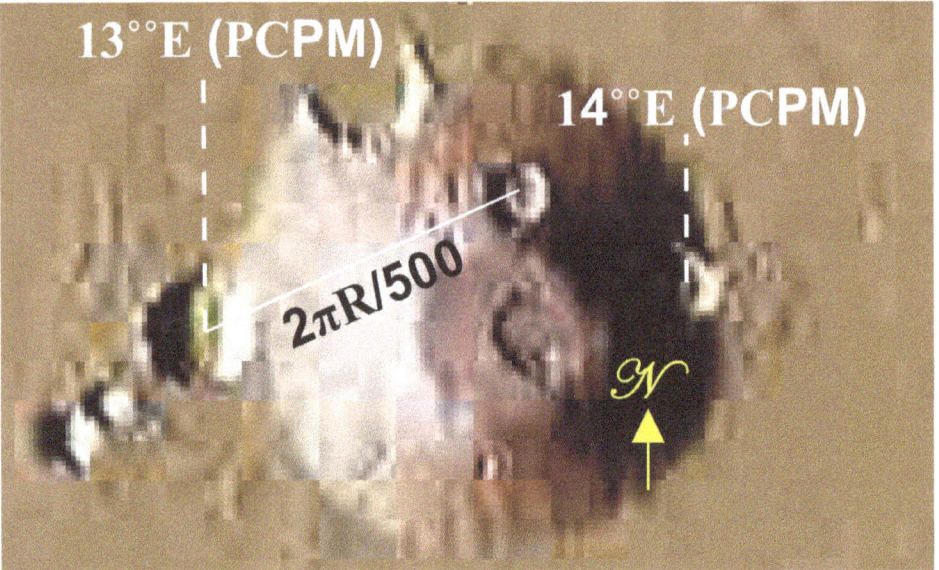

Fig. 13.7: *Craters associated with Uranius Tholus. The centres of the western and eastern craters are 1 big degree of longitude apart. The distance from the western crater to the northern crater is 1/500 of the planetary circumference. USGS Astrogeology.*

Tower PM), a numerical value close to 34 which is another Fibonacci number. The longitude is also 38.1804° W and 33.9381°° W with reference to the Sharonov Triangle PM. These numbers are very close to the matching formulae of $27\sqrt{2} = 38.1838$ and $24\sqrt{2} = 33.9411$. Note the use of the number 27 which is 1/4 the value of the angle between the star points of a pentagram. Finally, the longitude is 14.0017°° E (Pavonis Caldera PM) which is very close to the pure integer 14.

The western crater has the coordinates of 261.7984° E 26.1697° N. Its latitude is close to $10\varphi^2 = 26.1803$. In terms of sacred degrees it is 20.9358°°°N which is close to the matching formula of $8\varphi^2 = 20.9443$. The longitude with reference to the Pavonis Caldera PM is 13.0009°° E, a value very close to 13 which is a Fibonacci number. Hence, the western crater is almost exactly 1 big degree of longitude apart from the eastern crater.

The coordinates of the northern crater are 262.5374° E 26.4520° N. Its latitude is 34.5516° N (AMPL) which is close to $11\pi = 34.5575$. Its longitude is 30.9080° W (Sharonov Tower PM) which is close to $50/\varphi = 30.9017$. The importance of the northern crater becomes apparent in its relationship to the western crater. The distance between the 2 craters is 42.68 km which is 1/500 of the planet's equatorial circumference. The clockwise bearing angle between the 2 craters is 66.9206° which is about 10 minutes of a degree more than $108/\varphi = 66.7477°$. Note that 108° is the size of the angle between the star points of a pentagram.

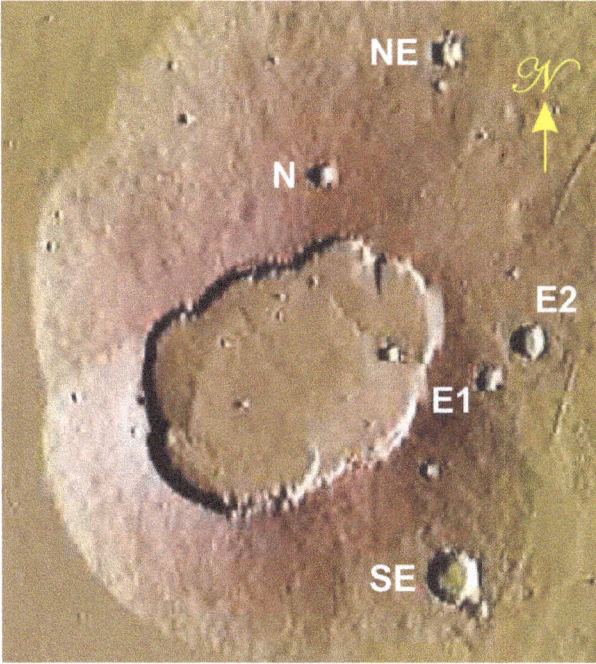

Fig. 13.8: *Craters on Uranius Mons labelled as northeast (NE), north (N), east #1 (E1), east #2 (E2) and southeast (SE). Interesting latitudes are obtained for each crater when the Arsia Mons Prime Latitude is used as reference. The NE crater is 29.0000°°° N (Arsia Mons PL). USGS Astrogeology.*

The distance of 23.26 km between the eastern and northern craters is close to R'/144 = 23.45 where R' is the northern polar radius. The counter-clockwise bearing angle between these 2 craters (62.0810°) is 8.66 minutes of a degree less than 44√2 = 62.2254°. The distance between the eastern and western craters is 60.15 km which is slightly longer than the distance of 59.80 km for a big degree of longitude at the latitude of the eastern crater. The reason for the discrepancy is that the western crater is at a slightly lower latitude than the eastern crater which increases the distance.

The craters associated with Uranius Tholus appear to be mostly honouring the pentagram (with the use of φ, 27, 108 and Fibonacci numbers) followed by the square (with the use of √2). They also seem to be honouring the planet with the distance of 1 big degree of longitude between the eastern and western craters, and with the distance of 1/500 of the planetary circumference between the western and northern craters.

Uranius Mons

There are several craters located on Uranius Mons but I chose to examine only the 5 largest. I have labeled these craters in Fig. 13.8 since they are without official names. All of these craters have interesting latitudes, but mostly with reference to the Arsia Mons Prime Latitude rather than to the Equator (Table 13.1). None of the interesting latitudes is in terms of

Table 13.1: *Latitudes of Uranius Mons craters from the Arsia Mons Prime Latitude and the Equator.*

Uranius Mons Crater	Latitude from Arsia Mons	Sacred Formula (Equation)	(Degrees)	Difference
NE	29.0000^{ooo}	29.0000^{ooo}	29.0000^{ooo}	0.0000
N	31.5888^{oo}	$\sqrt{3}/\pi$ big rad.	31.5888^{oo}	0.0000
N	28.4299^{ooo}	46/φ^{ooo}	28.4296^{ooo}	0.0003
E1	27.5000^{ooo}	27.5000^{ooo}	27.5000^{ooo}	0.0000
E2	30.7390^{oo}	19φ^{oo}	30.7426^{oo}	-0.0036
SE	26.5651^{ooo}	asin(1/$\sqrt{5}$)^{ooo}	26.5651^{ooo}	0.0000
Lat. from Eq.				
NE	22.5203^{ooo}	13$\sqrt{3}^{\text{ooo}}$	22.5167^{ooo}	0.0036
E1	21.0203^{ooo}	13φ^{ooo}	21.0344^{ooo}	-0.0141

regular degrees while 2 are in big degrees and 6 in sacred degrees. With reference to the Arsia Mons PL, the NE crater is exactly 29.0000^{ooo} N and the E1 crater is 27.5000^{ooo} N. Both these values are pure integer and pure integer and 1/2 numbers. The N crater is 31.5888^{oo} N (Arsia Mons PL) which is equal to $\sqrt{3}/\pi$ big radians converted to big degrees. The N crater is also 28.4299^{ooo} N (AMPL) which is very close to 46/φ = 28.4296. The E2 crater has a latitude with reference to the Arsia Mons PL of 30.7390^{oo} N which is close to 19φ = 30.7426. The SE crater is 26.5651^{ooo} N (Arsia Mons PL) which is equal to asin(1/$\sqrt{5}$). Finally, from the equator, the latitude of the NE crater is 22.5203^{ooo} N which is close to 13$\sqrt{3}$ = 22.5167, and the E1 crater is 21.0203^{ooo} N which is close to 13φ = 21.0344.

The Uranius Mons craters also show interesting longitudes with very small deviations from theoretical values (Table 13.2). Some of the craters have more than 1 longitude formula. The longitudes of the N and NE craters are 10φ^{ooo} E and 13$\varphi°$ E with reference to the Pavonis Caldera PM (note the match of the NE crater longitude with its latitude of 13$\sqrt{3}^{\text{ooo}}$ N in the use of the number 13). Matching formulae of 20$\varphi°$ W and 16φ^{ooo} W (Sharonov Triangle PM) were found for the longitude of the E2 crater. Other longitude formulae containing φ include the longitude of [47/φ]^{oo} W (Sharonov Tower PM) for the E1 crater and the longitude of [37φ^2]^{ooo} E (Elysium Mons PM) for the SE crater. Matching formulae of 10$\sqrt{3}^{\text{ooo}}$ E (Pavonis Mons PM) and 19$\sqrt{3}°$ W (Sharonov Tower PM) were found for the longitudes of the E2 and SE craters respectively. Longitude formulae containing the value of e are the longitude of 12e^{ooo} W (Sharonov

Table 13.2: *Longitudes of craters on Uranius Mons and their sacred formulae.*

Uranius Mons Crater	Longitude	Prime Meridian	Sacred Formula (Equation)	Sacred Formula (Degrees)	Difference
NE	$21.0443°$ E	PCPM	$13\varphi°$ E	$21.0344°$	0.0099
N	$16.1737^{\circ\circ\circ}$ E	PCPM	$10\varphi^{\circ\circ\circ}$ E	$16.1803^{\circ\circ\circ}$	-0.0066
N	$96.1737^{\circ\circ\circ}$ E	EMPM	$68\sqrt{2}^{\circ\circ\circ}$ E	$96.1665^{\circ\circ\circ}$	0.0072
E1	$29.0471^{\circ\circ}$ W	SToPM	$47/\varphi^{\circ\circ}$ W	$29.0476^{\circ\circ}$	-0.0005
E1	$32.6103^{\circ\circ\circ}$ W	STrPM	$12e^{\circ\circ\circ}$ W	$32.6194^{\circ\circ\circ}$	-0.0091
E2	$32.3535°$ W	STrPM	$20\varphi°$ W	$32.3607°$	-0.0072
E2	$25.8828^{\circ\circ\circ}$ W	STrPM	$16\varphi^{\circ\circ\circ}$ W	$25.8885^{\circ\circ\circ}$	-0.0057
E2	$17.3172^{\circ\circ\circ}$ E	PMPM	$10\sqrt{3}^{\circ\circ\circ}$ E	$17.3205^{\circ\circ\circ}$	-0.0033
SE	$21.0828°$ E	PCPM	$1/e$ (rad.) E	$21.0778°$	0.0050
SE	$32.9172°$ W	SToPM	$19\sqrt{3}°$ W	$32.9090°$	0.0082
SE	$96.0662^{\circ\circ\circ}$ E	DMPM	$(13e^2)^{\circ\circ\circ}$ E	$96.0577^{\circ\circ\circ}$	0.0085
SE	$96.8662^{\circ\circ\circ}$ E	EMPM	$(37\varphi^2)^{\circ\circ\circ}$ E	$96.8673^{\circ\circ\circ}$	-0.0011

Triangle PM) for the E1 crater, the longitude of $1/e$ radians = $21.0778°$ E (Pavonis Caldera PM) for the SE crater, and the longitude of $[13e^2]^{\circ\circ\circ}$ E (Dagger Midline PM), also for the SE crater. A single longitude formula containing the value of $\sqrt{2}$ is the formula of $68\sqrt{2}^{\circ\circ\circ}$ E (Elysium Mons PM) for the N crater. No formulae were found containing either π or $\sqrt{5}$.

When we look at the distances and bearing angles between the craters some interesting formulae turn up (Table 13.3). The distance from the SE crater to the NE crater is close to $R/(6\pi)$ km and the distance from the SE crater to the N crater is $\varphi R/(12\pi)$ km. Note the symmetry in the denominators. The distances from the SE crater to the E2 and E1 craters also show a symmetry, being 1/250 and 1/300 respectively of the equatorial circumference of the planet. The distance of 1/30 of the northern polar radius from the NE crater to the E1 crater is matched to the distance of 1/33 of the planetary equatorial radius from the NE crater to the E2 crater.

The clockwise bearing angle between the SE and E2 craters is $18°$ which is half the angle of a pentagram star point. Another interesting bearing angle is that of $7.5°$ between the E1 and NE craters. This is the angle of the groove lines which will be discussed in *Intelligent Mars III*. The bearing angle between the E2 and N craters is about 3.25 minutes of a degree smaller than $\text{acos}(1/\varphi)$ degrees. The remaining 3 bearing angles in Table 13.3 had more complicated formulae but were all within 2.24

Table 13.3: *Distances and bearing angles between Uranius Mons craters.*

| Uranius Mons Craters | Distance | | | |
| | Theoretical | | Actual | Difference |
	Formula	(km)	(km)	(km)
SE to NE	R/(6π)	180.17	180.42	0.26
SE to N	φR/(12π)	145.76	145.63	-0.13
SE to E2	2πR/250	85.36	85.69	0.34
SE to E1	2πR/300	70.71	70.44	-0.27
NE to E1	R'/30	112.54	112.10	-0.44
NE to E2	R/33	102.91	102.83	-0.08

| Uranius Mons Craters | Bearing Angle | | | |
| | Theoretical | | Actual | Difference |
	Formula	(°)	(°)	(°)
SE to NE	√3/e	0.6372	0.6482	0.0110
SE to N	asin(1/(√2√5))	18.4349	18.4169	-0.0180
SE to E2	-18	-18.0000	-17.9922	-0.0078
SE to E1	-atan(1/(π√3))	-10.4134	-10.4507	0.0373
E1 to NE	7.5	7.5000	7.5037	0.0037
E2 to N	acos(1/φ)	51.8273	51.7732	-0.0541

minutes of a degree of the theoretical formulae.

The presence of φ occurring so frequently in longitude formulae plus the bearing angle of 18° between the SE and E2 craters suggest a strong focus on the pentagram. However, the sacred distance formulae point to an honouring of the planet as well since 2 of the formulae are in terms of the planetary circumference and another 2 are simple functions of the equatorial and northern polar planetary radii.

Ceraunius Tholus

I examined the 2 largest craters on Ceraunius Tholus for coordinates, bearing angle and distance (Fig. 13.9). The centre of the crater on the north side of the mountain was found to be 29.0000°° north of the Arsia Mons PL. This is analogous to the NE crater on Uranius Mons which is 29°°° north of the Arsia Mons PL, the only difference being that the latitude of the crater on Ceraunius Tholus is in terms of big degrees instead of sacred degrees. The latitude of the Ceraunius Tholus north crater is also 32.6250° N (Arsia Mons PL) which is close to 12e = 32.6194.

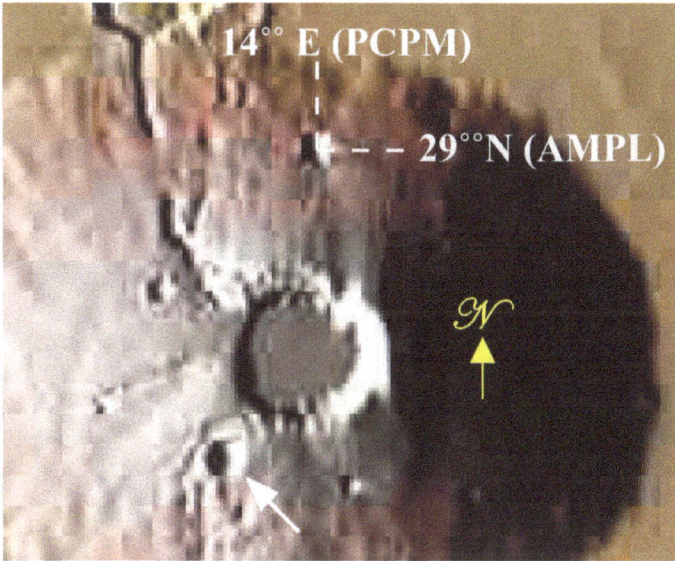

Fig. 13.9: *Ceraunius Tholus has 2 sizeable craters. The longitude of the north crater is almost exactly the same as the east crater on Uranius Tholus. The white arrow points to the south crater which is at a latitude of 1/e big radians. USGS Astrogeology*

The longitude of the Ceraunius Tholus north crater is 102.0000°° E (Dagger Midline PM), 14.0000°° E (Pavonis Caldera PM) and 34.0000°° W (Sharonov Tower PM). Remarkably, this is almost exactly the same longitude as the eastern crater on Uranius Tholus just to the north. These 2 craters could be used as a pointer to the north and south poles by overhead spacecraft similar to the Albor Tholus and Hecates Tholus pair of mountains. The longitude of the Ceraunius Tholus north crater is also 38.1823° W and 33.9398°° W (Sharonov Triangle PM). The numbers are very close to the values of $27\sqrt{2} = 38.1838$ and $24\sqrt{2} = 33.9411$. Note that 27 is 1/4 the size of the angle between the star points of a pentagram.

The latitude of the crater on the south side of Ceraunius Tholus is 1/e big radians = 21.0779°° N. This latitude is also 28.2775°° N (Arsia Mons PL) which is quite close to the value of $9\pi = 28.2743$, and it is 25.4498°°° N (Arsia Mons PL) which is quite close to $18\sqrt{2} = 25.4558$. Its longitude is 114.5405° E, 101.8138°° E and 91.6324°°° E (Dagger Peak PM). These numbers are close to $81\sqrt{2} = 114.5513$, $72\sqrt{2} = 101.8234$ and $35\varphi^2 = 91.6312$. Note that 72 is the size of the angles at the base of a pentagram star point.

The clockwise bearing angle between the 2 craters is 17.3482° which is close to $10\sqrt{3} = 17.3205$. The distance between the craters is 50.47 km for which I could not find a meaningful sacred formula. The sacred geometry of these 2 craters seems to emphasize the pentagram with the appearance of the numbers 9, 18, 27 and 72 in their coordinate values. These numbers are factors of the 3 angles of the pentagram. The coordinate values also emphasize the square with the repeated appearance of $\sqrt{2}$.

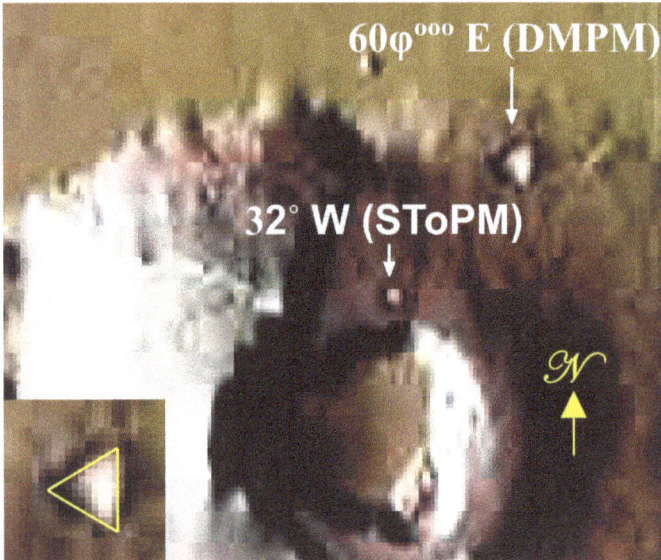

Fig. 13.10: *There are 2 craters on the northern side of Tharsis Tholus. The smaller crater near the caldera has a longitude of 32° W (Sharonov Tower PM). The crater near the northern edge of the mountain has a longitude of 60φ°°° E (Dagger Midline PM) and its interior can be fit to an equilateral triangle (see inset bottom left). USGS Astrogeology.*

Tharsis Tholus

There are 2 craters on the north side of Tharsis Tholus (Fig. 13.10). The smaller one just north of the caldera has a latitude of 13.8598° N which is close to $8\sqrt{3} = 13.8564$. Its longitude is 122.0002° E (Elysium Mons PM), 121.00002° E (Dagger Midline PM), 22.0002° E (Pavonis Caldera PM) and 31.9998° W (Sharonov Tower PM). The latter value, which is almost exactly the integer number 32, might refer to the pentagram (32 is the 5th power of 2) and the square ($32 = 8 \times 4$).

The larger crater which is near the northern edge of the mountain is at a latitude of 14.2274° N which has a numerical value close to $23/\varphi = 14.2148$. The longitude of this crater is 121.3546° E and 97.0839°°° E (Dagger Midline PM). These are interesting values since they are very close to $75\varphi = 121.3525$ and $60\varphi = 97.0820$. The integer 60 is the size of the angles of an equilateral triangle. Remarkably, the arrowhead shape in the interior of this crater can be well fit to an equilateral triangle as shown in the lower left inset in Fig. 13.10. The longitude of the crater is also 22.3546° E and 17.8837°°° E (Pavonis Caldera PM). This is an interesting pair of numbers since they are close to $10\sqrt{5} = 22.3607$ and $8\sqrt{5} = 17.8885$.

The distance between the 2 craters (29.84 km) is almost exactly one-half of the distance covered by a degree of latitude ($59.27/2 = 29.64$ km). The bearing angle between them is 43.0873° which is close to $\text{atan}(\varphi/\sqrt{3}) = 43.0508$. When we examine the sacred geometry themes of the coordinate and bearing angle formulae, we see an emphasis on the pentagram (φ, $\sqrt{5}$), the square ($8 = 2 \times 4$, $32 = 8 \times 4$) and the equilateral triangle (60, $\sqrt{3}$).

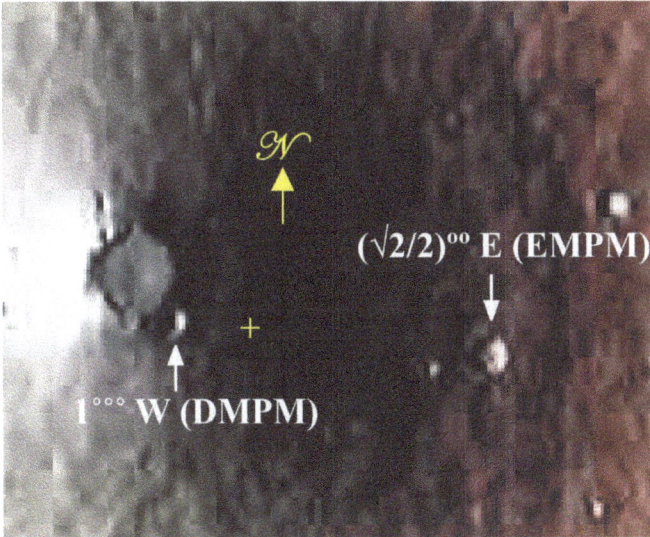

Fig. 13.11: *The small crater near the Elysium Mons Caldera is 1°°° W of the Dagger Midline PM. The eastern crater is 0.7071°° E (Elysium Mons PM) which is equal to (√2/2)°°. The yellow cross is the survey centre for Elysium Mons. USGS Astrogeology.*

Like the Ulysses Tholus craters, these 2 craters with the distance of 1/2 latitude degree between them may provide a means for spacecraft to determine altitude.

Elysium Mons

For Elysium Mons, I examined the small crater just southeast of the caldera and a larger crater near the eastern edge of the mountain (Fig. 13.11). The longitude of the centre of the small crater is 1.0002°°° W of the Dagger Midline PM. Its latitude is 32.6142° N (Arsia Mons PL) which is close to 12e = 32.6194°. Note that this is extremely close to the latitude of 32.6250° N (Arsia Mons PL) for the Ceraunius Tholus north crater which is almost exactly 116 degrees to the east.

The latitude of the centre of the larger crater to the east on Elysium Mons is 32.5072° N (Arsia Mons PL) which is very close to atan(√3/e) = 32.5047. Its latitude is also 26.0058°°° N (Aria Mons PL) which is quite close to a pure integer value of 26. The longitude of the crater is 0.7071°° E (Elysium Mons PM) which is equal to (√2/2)°°. This latter value may refer to the square shape of the Elysium Mons Caldera. The longitude of the eastern crater is also 0.8632° E (Crater Edge PM) which is very close to √3/2 = 0.8660. Thus the value of √3 is present in both the latitude and longitude of the eastern crater. The value of e is present in the latitude coordinates of both craters.

The distance between the 2 craters is 56.77 km which is almost equal to R/60 = 56.60 km. The clockwise bearing angle from the crater near the caldera to the eastern crater is 96.4139° which is very close to 156/φ =

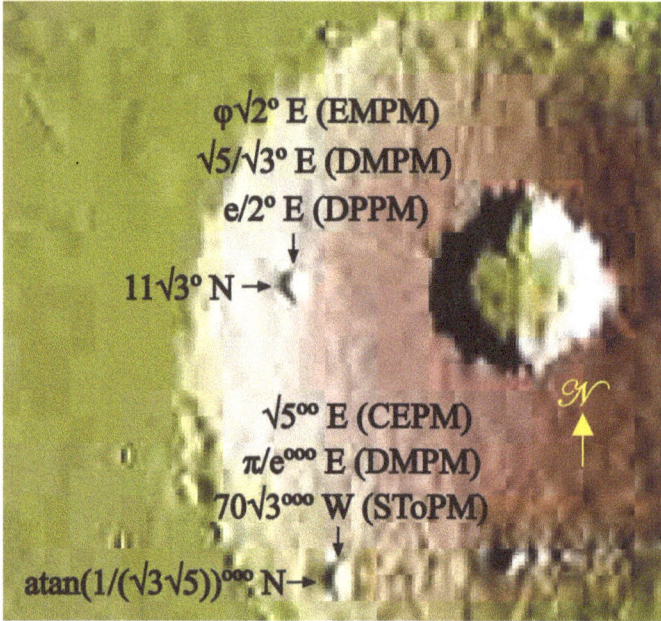

Fig. 13.12: *A western crater and a southwestern crater are located on the west half of Albor Tholus. The western crater has a latitude of 11√3° N and has several meaningful longitude formulae with reference to various prime meridians. Only some are shown. The southwestern crater is positioned at a latitude of atan(1/(√3√5))°°° N and also has several meaningful longitude formulae. USGS Astrogeology.*

96.4133. Besides the emphasis on e, √3 and √2, these craters also seem to emphasize the integers of 12 and 13. The number 156 is the product of 12 and 13, and 26 is twice 13. Also, as stated above, the latitude of the crater near the caldera is 12e° N (Arsia Mons PL).

Albor Tholus

The centre (149.4639° E 19.0521° N) of the crater west of the Albor Tholus Caldera (Fig. 13.12) has a latitude which is very close to 11√3 = 19.0526° N. It is very rich in meaningful longitudes. It is 2.2915° E and 2.0369°° E (Elysium Mons PM). These numbers are close to φ√2 =2.2882 and 3e/4 = 2.0387. With reference to the Dagger Midline PM, the longitude is 1.2915° E and 1.1480°° E. The first number is close to 3√3/4 = 1.2990 and very close to √5/√3 = 1.2910. The second number is close to φ/√2 = 1.1441. With reference to the Dagger Peak PM, the longitude is 1.3592° E which is very close to e/2 = 1.3591.

The second crater (149.6180° E 18.0969° N) that I examined on Albor Tholus is located near the southwest edge of the mountain (Fig. 13.12). It has a latitude of 14.4775°°° N which is equal to atan(1/(√3√5)). Like the western crater, the southwest crater is rich in meaningful longitudes. It is 2.4456° E and 1.9565°°° E (Elysium Mons PM). These numbers are close to √2√3 = 2.4495 and 3φ²/4 = 1.9635. With reference to the Dagger Midline PM it is 1.1565°°° E which is very close to π/e = 1.1557. With reference to

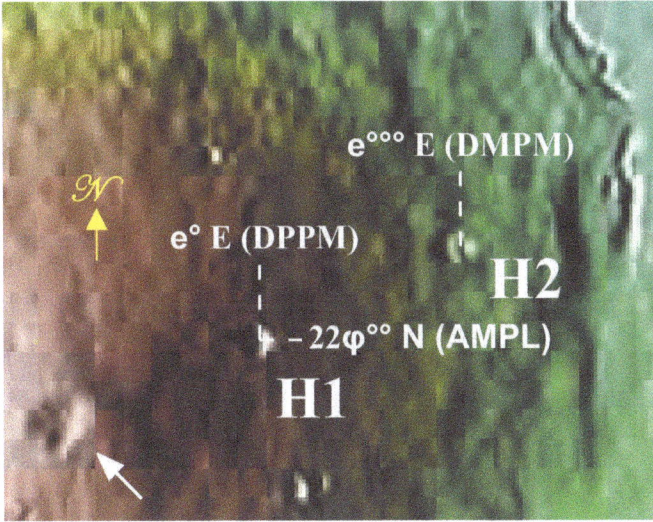

Fig. 13.13: *Two of the craters on Hecates Tholus are located at e degrees of longitude but these longitude coordinates differ in degree systems and prime meridians. The H1 crater has a latitude of 22φ°° N (AMPL). The white arrow points to the Hecates Tholus Caldera. USGS Astrogeology.*

the Dagger Peak PM it is 1.2106°°° E which is very close to 3φ/4 = 1.2135. With reference to the Crater Edge PM it is 2.2340°° E which is very close to √5 = 2.2361. With reference to the Sharonov Tower PM it is 121.2435°°° W which is almost exactly 70√3 = 121.2436.

The distance between the 2 craters is 57.28 km which is close to R/60 = 56.60 km. This is the same distance formula between the 2 craters on Elysium Mons. The counterclockwise bearing angle from the southwest crater to the west crater is 8.6939° which is close to the sacred formula of 5√3 = 8.6603.

With the coordinate sacred formulae, all the primary irrational numbers (π, φ, e, √2, √3 and √5) are present in the coding of these 2 craters on Albor Tholus. However, the greatest emphasis is on √3 and φ which represent the equilateral triangle (the height of an equilateral triangle = √3 times 1/2 the side length) and the pentagram respectively. This is supported by the sacred formula of 5√3 for the bearing angle between the 2 craters where 5 would refer to the 5 star points of the pentagram and √3 would refer to the equilateral triangle. The distance formula of R/60 suggests the 60 degree angles of the equilateral triangle as well.

Hecates Tholus

The 2 largest craters on the east side of Hecates Tholus (Fig. 13.13) are very remarkable in that they both have longitudes which are extremely close to e degrees. However, the longitudes differ in the degree system used and the prime meridian to which they refer. The longitude of the crater (labeled H1) which is closer to the caldera is referenced to the

Dagger Peak PM and is in regular degrees (2.7188° E vs. 2.7183 for e). The longitude of the crater (labeled H2) near the northeast edge of the mountain is referenced to the Dagger Midline PM and is in sacred degrees (2.7191°°° E vs. 2.7183 for e). The latitude of the H1 crater is 35.5924°° N (AMPL) which is close to $22\varphi = 35.5968$, whereas the H2 crater has a latitude of 35.8352°° N (AMPL) which is close to $58/\varphi = 35.8460$.

Both of these craters have several more sacred coordinate formulae for their longitudes. The H1 crater has a longitude of 3.2454°° E (Elysium Mons PM) which is close to $2\varphi = 3.2361$. It also has a longitude of 2.1209°°° E (Dagger Midline PM) which is very close to $3/\sqrt{2} = 2.1213$. The H2 crater is 4.3989° E (Elysium Mons PM) which is very close to $e\varphi = 4.3983$. It is also 3.4666° E (Dagger Peak PM) and 4.4666° E (Crater Edge PM). These numbers are close to $2\sqrt{3} = 3.4641$ and $2\sqrt{5} = 4.4721$.

The distance between the 2 craters is 40.90 km for which I could not find a meaningful formula. The bearing angle between them is 66.6792° which is about 4 minutes of a degree less than the sacred formula $108/\varphi = 66.7477°$. Note that 108 is the size in degrees of the angles between the star points of a pentagram. It should be mentioned here that the caldera on Hecates Tholus also has interesting longitude values. It is 2.0032° E (Dagger Peak PM) and 3.0032° E (Crater Edge PM). The integers of 2 and 3 play a very important role in numerology and sacred geometry.

Although all the primary irrational numbers except π are present in the coding of these 2 craters on Hecates Tholus, the primary emphasis seems to be on the number e and on the pentagram (with φ, $\sqrt{5}$, 108). The number e also probably references the pentagram.

Conclusion

It turns out that many of the craters which are present on the major mountains of Mars are not haphazardly located at all. This strongly questions a meteoric origin and points to engineering by an advanced civilization. Of the 25 craters which I examined on 10 mountains, I found that every one of them had meaningful latitudes and longitudes. What is interesting is that for the meaningful latitudes, 16 of the 25 craters use the Arsia Mons PL and 14 use the equator (5 use both). This is a good demonstration of the importance of Arsia Mons as a reference point for latitude in addition to the equator. No less than 4 of the values using the Arsia Mons PL were pure integers in the sacred degree system. Assuming an accuracy of ±0.01 degrees, this result has only about a 1 in 692 chance of occurring with 25 craters under random conditions, so it is very unlikely to have been natural. Another 2 pure integer latitudes using the Arsia Mons PL are also present, 1 in regular degrees (which also worked out to an integer in sacred degrees for one of the 4 craters above) and the other in big degrees. No pure integer latitudes with a deviation of less than ±0.01 degrees were found with the equator as reference.

Integer longitudes (within ±0.005 degrees) were found for 7 of the craters, 1 in terms of regular degrees, 3 in big degrees and 3 in sacred degrees. The integer longitudes of 2 of the craters associated with Uranius Tholus differed by 1 big degree and one of these had the same integer longitude as a crater on Ceraunius Tholus. Any theory suggesting that these craters were used to mark out some sort of coordinate gridline system for overhead spacecraft, however, does not seem plausible since only 9 of the 25 craters had an integer longitude or latitude or both, and the integers were scattered amongst the 3 different degree systems.

There is the remarkable finding of 2 craters on Hecates Tholus with a longitude of e degrees, one in terms of regular degrees (DPPM) and the other in terms of sacred degrees (DMPM). For Albor Tholus, the western crater has a longitude of $e/2°$ E (DPPM) and the southwestern crater has a longitude of $\pi/e°°°$ E (DMPM). For Elysium Mons the crater close to the caldera has a latitude of $12e°$ (AMPL) and the eastern crater has a latitude of $\text{atan}(\sqrt{3}/e)°$ (AMPL). So all of the craters examined for these 3 mountains have e appearing in their coordinate values, twice by itself and 4 times in association with other numbers. This is highly unlikely to occur by chance alone.

Many of the distances and bearing angles between craters were remarkable as well. Craters on Ulysses Tholus mark out the length of a degree of latitude and craters on Tharsis Tholus mark out the half-length of a degree of latitude giving convenient calibration lines which could be used by overhead spacecraft to determine altitude. Although all of the irrational numbers are present in sacred formulae for coordinates and in bearing and distance formulae, the number φ occurs 43 times compared to 18 – 20 times each for π, e, $\sqrt{2}$ and $\sqrt{3}$, and 14 times for $\sqrt{5}$. This indicates that the craters are principally focused on representing the pentagram, especially since the integers of 5, 10, 9, 18, 36, 72 and 108 occur 26 times in the sacred formulae and in pure integer coordinates. These integers represent the number of star points and isosceles triangles as well as the angle values in a pentagram. The formulae and integers also represent the square with the occurrence of $\sqrt{2}$, 8, 16 and 32, and the equilateral triangle with the occurrence of $\sqrt{3}$, 60 and 30, but these were found to occur with less than half the frequency of numbers related to the pentagram in both instances.

14

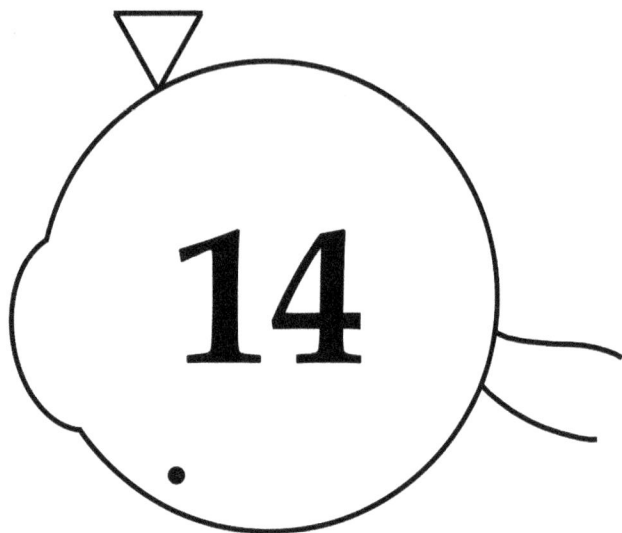

Craters as Teachers

This book has examined many features of the craters on Mars including coordinates, shapes, sizes, bearing angles and distances from each other and other sites, internal structures and linear segments. What lessons then have we learned from them?

Perhaps the most important lesson is that a huge number of the craters are artificial in nature and cannot be the result of random meteorite strikes as current astronomical science teaches us. For one thing, many craters are not round as would be expected if they were made by an asteroid or comet striking the planet from outer space. This simple fact seems to either escape the attention of scientists (most likely due to the affliction of inattentional blindness which I have often been affected with myself), or it gets explained away by rather dubious theories. One of the most obvious deviations from circular is the 6E3S Crater which has not even been given an official name. Other craters have polygonal shapes which are rather quite obvious once you catch on to the concept, but have generally been somewhat camouflaged by the architects to escape such attention. Even the sizes of craters give away their artificiality but it takes a considerable amount of effort to detect this. Many craters come in standard sizes rather than follow a distribution of sizes based on randomness, and the polygonal craters are sized according to chromatic scale music intervals. So instead of being evidence of a bleak lifeless planet, once you understand the artificial nature of many of the craters, they point to a very advanced civilization having existed on Mars in the distant past.

The next most important lesson is that every artificially constructed crater appears to serve some function or other, and it is this function that often reveals their artificiality. They don't just sit anywhere nor are the shapes of the noncircular craters simply haphazard. A very important function is to create significant numbers (integers, integer and 1/2 values, or primary irrational numbers) or very meaningful sacred geometry formulae with the coordinate values of their centres (or of their vertices in the case of polygonal craters). These numbers do not get revealed by NASA coordinates other than for some latitude values. They require reference to special prime meridians and/or often the Arsia Mons Prime Latitude in order for them to come to life. As well, there is plenty of evidence that big degrees (1.125 regular degrees) and sacred degrees (1.25 regular degrees) were used in addition to the degrees of a 360 degree system. Distances and bearing angles between craters and other sites can often be expressed as sacred geometry formulae which convey their own special meaning. Certain craters, as discussed in *Intelligent Mars I*, serve the function of survey points to calculate the location of the centres of mountains by a triangulation procedure. Several craters could serve a compass function. The square craters are most often oriented so that their vertices point in the cardinal directions. Pentagon craters always have one of their vertices pointing in a cardinal direction, most often to the east. Hexagon craters most often have 2 vertices pointing NS or 2 vertices pointing EW. Like Albor Tholus and Hecates Tholus, 2 craters or a crater and a caldera can team up to point to the planetary poles (e.g., the north crater on Ceraunius Tholus together with the eastern crater on Uranius Tholus).

A third lesson to be learned from the craters is something about the nature and purpose of the sacred geometry of the planetary architecture, and by extension, something about the values and culture of the civilization that created it. The most prominent theme appears to be the pentagram which the craters point to by certain integers and irrational numbers reflecting its angles and internal structural relationships. The integers refer to the pentagram angles or their halves and quarters (108, 54, 27, 72, 36, 18, 9). As discussed in *Intelligent Mars I*, the irrational numbers φ, √5 and e can be found in the pentagram structure and therefore can be used to represent the pentagram itself. Other prominent themes are the equilateral triangle, the square and the circle. The equilateral triangle is pointed to by the integers 60 and 30 (1/2 of 60) which refer to the internal angles, and by √3 which is a factor in its height. The square is pointed to by the number 4 and its binary multiples (e.g., 8, 16) and by √2 which is a factor in its diagonal length. The circle, represented by π, encompasses the equilateral triangle, the square and the pentagram, and is the underlying shape of all the craters including those having the shape of a regular polygon. But why these 4 geometric shapes? What is so important about them to the Martian architects? In my analysis of the pentagram I came to the conclusion that it represents Divine creation, fertility and growth. In ancient traditions, the equilateral triangle represents the Divine, completion and unity, and the square represents the stability and the natural order of things. The circle encompasses all of these meanings and also represents infinity and timelessness. In other words these 4 geometric shapes have to do with the creation and structure of the universe by an infinite Being. Since there is no direct indication of such a Being in the crater or mountain architecture it would appear that the Martians worshipped a Divine Being that has no shape or form.

One of the most astounding findings with the polygon craters is that the Martians understood the chromatic scale and probably had a system of music not all that different from some of the systems used on this planet both today and in ancient times. This leads me to suspect that much of what we regard as an evolutionary development of the arts and sciences by ancient cultures and by talented individuals within those cultures is actually knowledge passed down over a period of billions of years from extraterrestrial cultures inhabiting Mars or others planets in the cosmos. Concepts such as a prime meridian and a 360° coordinate system, geometric shapes, music, sacred geometry and irrational numbers may have been inherited in part rather than developed anew.

The discovery that the Elorza and Sharonov craters are shaped very much like the human eye gives further impetus to the theory presented in *Intelligent Mars I* that the Martians may be our ancestors or may have derived from a humanoid species common to our own ancestry. It was hypothesized that the bisected isosceles triangle formed by the giant mountains may have represented a body shape in which the length of the horizontally extended arms is equal to body height, and thus, similar to our own structure. It was also suggested that Da Vinci's Vitruvian Man is a copy of a virtual Vitruvian Martian which can be assembled from the giant mountains on the Tharsis Rise and the Pentagram Pyramid.

Many of the structures could only have been viewed from an elevation of many kilometers to make any sense of them (e.g., Janssen's Wheel, the distance of 1 latitude degree between the craters on Ulysses Mons, the eye of Sharonov with a triangle to mark a prime meridian). This suggests that the Martians were capable not only of air flight, but also of space travel. The creation of massive craters (some exceeding 100 km in diameter) with great precision indicates a high degree of engineering sophistication. Whether they were constructed by excavation, explosives or by other technologies such as energy beams, enormous amounts of highly controlled energy would have been required.

We have also been introduced to the concept of auxiliary craters. These craters are generally smaller craters sitting close to the edge, either inside or outside, of larger craters, to assist in the fitting of a geometric figure by providing an edge or centre for alignment. In addition, polygonal and other craters sometimes have linear edges near or at their perimeters which can be used to fit geometric figures or shapes (e.g., arrowheads). These edges can appear as notches or steps in the perimeter, or they can provide a linear guide over several kilometers. I have learned that any deviation from what you would normally expect in a crater has a great likelihood of serving as a marking for a geometric shape to fit. The shape then points to meaningful coordinates with its vertices. There are possibly many other functions to such deviations yet to be discovered.

From what was learned with the study of the polygonal-shaped craters, the concept of participatory sacred geometry was formulated. The polygonal-shaped craters provide only a part of the information required to create the geometric shape which fits them. The observer has to fill in the blanks in order to arrive at the final product. We have seen this in *Intelligent Mars I* where only the 2 northern star points of the Pentagram Pyramid and the surveyed positions of Olympus Mons and Pavonis Mons are provided for the observer to construct the circle, equilateral triangle and pentagram fitting the Vitruvian Martian. Hence, the observer has to play an active role to create the final product from given starting points.

A final lesson is that the craters bring a huge validation for big degrees and sacred degrees as well as for 8 prime meridians and a second prime latitude. The best way in which they do so is by having coordinates in terms of a single primary irrational number. This is especially important in establishing the validity of big and sacred degrees since the irrational numbers cannot be derived from integer or integer plus $1/2$ or $1/4$ values in regular degrees simply by dividing by 1.125 or 1.25. The numbers φ, π, e, $\sqrt{2}$, $\sqrt{3}$ and $\sqrt{5}$ are all extremely meaningful in a sacred geometry which celebrates these mystical constants as fundamental to the creation of the universe. In Table 14.1, all of these constants appear as coordinate values for the centres (except for the east and west vertices of the pentagon fitting the E966 crater) of craters and the Albor Tholus Caldera. The value of π appears as a coordinate for all 3 sizes of degree. The value of φ occurs for regular and sacred degrees, and its inverse Φ occurs for both big and sacred degrees. The value e occurs for regular and sacred degrees, while the value of $\sqrt{5}$ appears for regular and big degrees. The value of $\sqrt{3}$ appears for big and sacred degrees and the value of $\sqrt{2}$ appears for sacred degrees. For latitudes, 6 irrational numbers use the equator and 2 use the Arsia Mons Prime Latitude. For longitudes, all of the prime meridians are

Table 14.1: *Craters and one caldera with a coordinate having the value of a single primary irrational number.*

Site	Irrational Number	Prime Meridian or Prime Latitude
nn65 crater	$\varphi°$ S	Equator
nn37 crater	Φ^{oo} N	Equator
Janssen Crater	$e°$ N	Equator
nn23 crater	e^{ooo} S	AMPL
Ulysses Tholus N Crater	π^{oo} N	Equator
nn35 crater	$\sqrt{5}^{oo}$ S	Equator
Wallula Crater	$\sqrt{3}^{ooo}$ S	AMPL
E966 crater	$\sqrt{2}^{ooo}$ N	Equator
Albor Tholus Caldera	$\pi°$ E	CEPM
Albor Tholus SW crater	$\sqrt{5}^{oo}$ E	CEPM
Hecates Tholus H1 crater	$e°$ E	DPPM
Hecates Tholus H2 crater	e^{ooo} E	DMPM
AscSC2 crater	φ^{ooo} E	PMPM
AscSC2 crater	$\sqrt{3}^{oo}$ E	PCPM
Uranius Mons SE crater	1/e radians E	PCPM
Rauch Crater	Φ^{ooo} E	STrPM
Wallula Crater	π^{ooo} E	STrPM
ET638 crater	$\sqrt{5}°$ E	SToPM

used except Elysium Mons. However, Albor Tholus and Hecates Tholus are $\pi°$ E (EMPM) and the eastern crater on Elysium Mons is exactly $\sqrt{2}/2^{oo}$ E (EMPM) so these sites are sufficient to establish Elysium Mons as a prime meridian. In summary, with the help of data obtained from the study of craters, all the prime meridians and degree systems are now well established and this suprasystem will continue to be used in the final volume of the *Intelligent Mars* series.

There is one crater that I did not mention in this book even though I spent over a year trying to analyze it. It is an unnamed crater which I have reserved for *Intelligent Mars III* because it requires several chapters to adequately deal with it. This seemingly insignificant crater might be the most important structure on the planet and goes a long way towards explaining the intent of the architects. The third book also examines very large scale artifacts and a possible replica of the individual who may have been the genius behind the sacred geometry of the site layout on Mars.

***

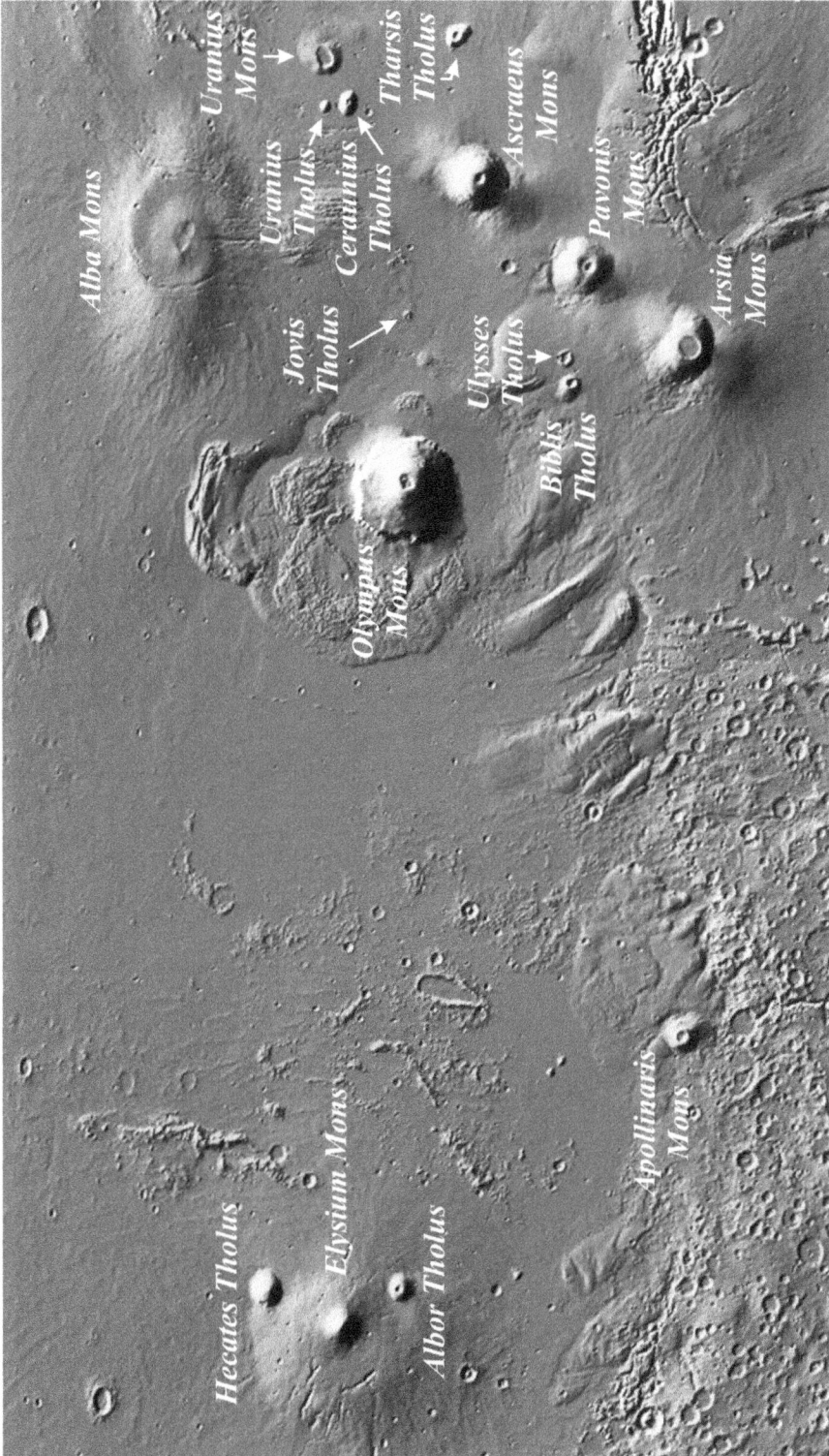

The mountains of Mars discussed in this book. Pavonis Mons lies just north of the equator, and Hecates Tholus just north of 30° N. This image covers about 138 degrees of longitude. MOLA Science Team Courtesy NASA/JPL-Caltech.

www.ingramcontent.com/pod-product-compliance
Lightning Source LLC
Chambersburg PA
CBHW042310210326
41598CB00041B/7337